Tomato Gardening & Cooking

BY JACQUELINE HÉRITEAU

GROSSET
GOOD LIFE
BOOKS

PUBLISHERS • GROSSET & DUNLAP • NEW YORK

Acknowledgments

Cover photograph by Mort Engel
Drawings by Peter Kalberkamp

My thanks to Grosset & Dunlap editor Lee Schryver, who brought to my attention this book all about tomatoes. I'm grateful because I love tomatoes and I've never yet had a chance to tell *everything* I know about them. My thanks, also, to Peter Kalberkamp who delightfully portrays the many ways tomatoes can be staked. To the commercial growers and producers of seed catalogs who helped us by providing information about exciting upcoming varieties of tomatoes, my thanks. My thanks to the U.S. Department of Agriculture for photographs of the basket-growing of tomatoes, and to Dr. H. Marc Cathey, horticulturist with the U.S.D.A.'s Research Service, my gratitude for this and many past favors. My gratitude to Evelyn H. Johnson, Ph.D., specialist in food and nutrition with the U.S. Department of Agriculture Extension Service in Washington, D.C., for her technical advice.

In addition, my thanks and appreciation to the following for their permission to use the photographs in this book: Bernard L. Lewis, Inc.: p. 55; Burpee Seeds: p. 12 top, p. 42; Dudley-Anderson-Yutzy, Inc.: p. 47, p. 50, p. 51, p. 52, p. 53; Earl May Seed and Nursery Co.: p. 13 top, p. 13 bottom, p. 14, p. 15, p. 40; Derek Fell and Leopold Klein: p. 34 top, p. 34 bottom; George J. Ball, Inc.: p. 16 top; Elvin McDonald: p. 16 bottom, p. 46; National Garden Bureau: p. 10; United States Department of Agriculture (Agricultural Research Service): p. 43 top, p. 43 middle, p. 43 bottom, p. 44; United States Department of Agriculture (Bureau of Plant Industry): p. 8, p. 12 bottom; United States Department of Agriculture (photo by Murray Lemmon): p. 28 top, p. 28 bottom, p. 29 top, p. 29 bottom; Westinghouse: p. 30.

Contents

Part I
Growing Tomatoes

Tomatoes tied to a gazebo. Flowering vines may be more traditional for a gazebo, but the "love apple" seems appropriate in a trysting place. And at the same time, marigolds surrounding your tomato patch will help to keep out slugs.

1
Everything about Tomatoes

The most popular of all food plants for the home garden is the tomato. As for eating, four out of five adults like or prefer tomatoes to any other vegetables, along with corn and potatoes, and two out of three of the younger members of the family, according to U.S. Department of Agriculture researchers.

Whether it is red, yellow, orange, or pink, whether you call it a fruit or a vegetable, whether you grow it on a fire-escape landing or in a suburban garden, whether you put it up in a jar, or freeze it juiced, sauced, whole, crushed, puréed, it's the single vegetable (oops—fruit) you'll find most rewarding.

This book is all about, and only about, tomatoes—what they are really like, the varieties, flavors, colors; what they grow in; how to grow them anywhere (even in a hanging basket); harvesting and storing them; canning, freezing, and cooking them.

A vine-ripened tomato is a thing of glory. It tastes very good, doesn't cost very much to raise, it grows easily, and has lots of vitamins A and C. And everybody loves it.

According to a Gallup organization, reported *The Christian Science Monitor* in late January 1975, almost half of all U.S. households grew some kind of garden in 1974, an increase of 8 million over three years. And tomatoes are far more saving in costs and easier to grow than the next two most popular vegetables—corn and potatoes.

The tomato has a rather disreputable history. People used to think it was a poison, and then they decided it was an aphrodisiac, and called it "love apple," *pomme d'amour*. Of course, it's neither. Most of all, it is the mainstay of the American summer salad and the Italian spaghetti sauce, among dozens of other things.

It is the most popular of all food plants in our country, not only because it's so versatile (look at the recipe sections of this book to see the range of ways it can be used) but also because it grows almost anywhere.

The wild Peruvian tomato, ancestor of many full-sized varieties of today, held by a U.S.D.A. researcher. On the ground in front of him are three Marglobe tomatoes (left) and three of the new Pan-American fruit (right).

When you're really planning to go into the growing of something like a tomato plant, it helps to understand something of its background.

The tomato has lots of relatives, all those plants belonging to the family called Solanaceae, the Nightshade. Sometimes called the potato family, it includes plants that are vegetables, as well as tobacco and some narcotics. Browalia is related to tomatoes, as are red pepper, Capsicum, Datura (which has a bad reputation, too), Nicotiana (the tobacco plant), the petunia, and the eggplant, to name a few. There's a wild vine also, "deadly nightshade," which has poisonous blossoms and leaves like tomato leaves. That's probably where the tomato's reputation as a Borgia originated.

It is important to rotate location to avoid wilt or other diseases. That is, not to plant tomatoes this season where tomatoes grew last season. This rule applies not only to tomatoes, but generally to other members of the tomato family. Knowing the tomato's close family ties, you can also know that it isn't good gardening practice to follow eggplants, tobacco, or petunias with tomatoes, as they are susceptible to the same diseases.

Although some people think the tomato belongs to Italy, because tomatoes turn up so often in Italian cooking, actually it is one of the few plants we use today that originated in America. All the forms of tomato are derived from a wild species that began in the section of South America that is today Peru, Ecuador, and Bolivia. In Mexico, long before Columbus, natives were cultivating forms of tomatoes, and it is generally believed the tomato took that route to Europe.

The first record we have of a description of the tomato appears in Italy in 1554. The name was "golden apple," *pomi d'oro,* close to today's Italian name for tomato, *pomodoro.* Tomato scholars argue that if the tomato the sixteenth-century Italians were describing was red, they wouldn't have called it golden, so it seems likely the first forms were yellow. Like the lovely little yellow pear and plum tomatoes offered in catalogs today.

From Italy *pomi d'oro* spread to French, Spanish, and English gardens, in both yellow and scarlet forms, but they were grown as horticultural curiosities. The name "love apple" dates from this period, and sometime, when you are pruning tomatoes, or repotting them, meditate on how many potentially frustrated lovers of the sixteenth century tried tomatoes on their sweethearts with no—or maybe some—effect. Tomatoes are not, of course, an aphrodisiac—except perhaps to people who are really very crazy about tomatoes.

By the mid-1700s, Europeans were eating tomatoes, but the Americans who grew them were very conservative and wouldn't touch them as food. Thomas Jefferson in 1781 left us one of the first records in our country of growing tomatoes, still not considered food until some adventurous Louisianians tried them, in 1812. Salem, Massachusetts, in 1802 condemned them as food—wouldn't you know—but by 1835, the editor of the *Maine Farmer* said they were "useful articles of diet." Until the turn of this century many people still believed tomatoes to be poisonous.

Once all that was settled, a mild controversy developed over whether the tomato was a fruit or a vegetable. Most of us still stumble over that one. Botanically, the tomato is a fruit. However, in the United States, for purposes of trade, it is called a vegetable. It was so classified in a decision by the U.S. Supreme Court in 1893. The reason the Court ruled it a vegetable is that it is used most often in the main part of a meal, which is when vegetables appear. That's not a very good reason for classifying it as a vegetable—or anyway, any horticulturist will tell you that that's not a good reason.

What tomatoes look like is less easy to be clear about. Though most of us expect them to be round and red, tomatoes come in many shapes, colors, and sizes that grow in quantity every year as commercial seedsmen work to develop new and exciting varieties of this all-time favorite.

'Big Early Hybrid,' a large red-fruited variety.

2
Finding the Right Tomato

It isn't easy. There are many, many tomato varieties and strains of varieties offered by growers. They do divide into groups, however, that can be related to what you want from the plants, and what they are going to need from you.

Generally speaking, the groups include size differences; there are miniature fruiting types that come in red, pink, and yellow; there are beefsteak types, big meaty tomatoes with less juice than those known as standard size; there are pink-fruited tomatoes that come in different sizes and ripen at different times; there are yellow tomatoes shaped like small pears or plums; there are large orange tomatoes (called yellow popularly), touted as low in acid; there are white tomatoes which I don't understand at all; there are tomatoes for canning and for making tomato paste; there are tomatoes that climb more readily than others; there are disease-resistant strains in almost all popular varieties. And then there's the tomato that isn't a tomato—the ground cherry.

What all this means to you is that you really have to think a little before you pick a tomato for your garden. Or else, you can pick whatever tomato the local garden center recommends and let it go at that—for the first few seasons, anyway. Local offerings usually grow well in local soils, are suited to the needs of most local growers, be they patio or penthouse gardeners, and are resistant to any tomato diseases prevalent in the area.

Early, Midseason, and Late Varieties

Since almost all tomatoes offered are described as belonging to one or the other of these categories, there's no point in listing them for you. However, the titles do bear understanding.

Early tomatoes are tomatoes, whatever size, shape, color, or growing habit, that will mature sooner than midseason or late varieties. Generally, they stand up to more cold early in the season than other plants. Sometimes they are especially popular strains

'Early Girl Hybrid Tomato.'

Various types of smaller tomatoes, ranging from currant and cherry through plum, pepper, and egg-shaped varieties.

that growers have worked to develop into types that mature quickly.

In planning your tomato garden, you will benefit from having a few early types to give crops while the longer-to-mature varieties are developing. These are mostly of standard size.

While early large varieties mature sooner than later varieties, few really big tomatoes are truly early—only earlier than other big varieties as a rule. It takes longer for a great big fruit to grow up and turn color.

Midseason fruits are those ready in the middle of the growing season, obviously enough, and usually these have stopped producing in long season climes by the time the late varieties come in, in late summer and early fall. Late varieties usually stand up better than others to fall frosts, have large, very tasty fruit when mature, and produce for as long as the vines are living. Of course, if the "suckers" are removed (see page 39), the strength of the vine will go into early maturity of the fruit, rather than into a lot of extra leaves and a super-crop of late-developing tomatoes.

Miniatures

Bred from the little currant tomato, varieties of this type are sometimes vines that will grow up trellises, sometimes dwarf bushes that need no staking. Some of the plants produce fruit in clusters, others produce fruit individually matured on branch tips. In other words, there's a small one for almost every purpose.

Among the most-grown of the miniature types are 'Tiny Tim' and 'Small Fry,' which mature in between 50 and 55 days, and are offered by most growers. 'Small Fry' is disease resistant in the varieties offered by some of the seedsmen. 'Pixie' is another miniature offered by seedsmen, as suited to window boxes and for ornamental borders. It matures in 52 days, and bears fruit 1¾ inches across. 'Early Salad Hybrid' is one of the smallest in size with vines 6 to 8 inches tall, bearing 1½-inch fruits. 'Small Fry' is the most vigorous grower of this group.

Beefsteak Tomatoes

Variations on the beefsteak theme proliferate in all the catalogs. Introduced a few years ago, these huge, meaty tomatoes became the darling of most growers. There's more to eat in these tomatoes and less to drink. They are excellent for cooking, less suited for the making of juice. They look impressive in a bowl, on the vine, sliced on a plate—and grow on largish plants that require really solid staking.

Among catalog offerings are beefsteak types with names like 'Oxheart,' 85 days; 'Beefeater Hybrid,' 60 days, a type reputed to give 2-pound size fruit early. Not all beefsteak tomatoes are giants. Some bear standard-size fruit on bushy plants and you can't judge by the days to maturity what size the fruit will be.

Pink-fruited Tomatoes

These are offered by most growers in varieties that are more or less midseason in maturing, or even rather late. They are said to be acid free, but the flavor is similar to that of the red tomatoes.

Among varieties offered, are 'Burpees Glove,' 80 days to maturity; 'Oxheart,' 86 days; 'Ponderosa,' a pink beefsteak type, 83 days; 'Crackproof Pink,' 78 days; 'Early Detroit,' 78 days; 'McMullen,' 74 days; 'Livingston Globe,' 75 days; and 'Kurihana,' 90 days, a Japanese tomato now being sold by Nichols. 'Early Pink,' is a miniature offered for use in patio gardens, window boxes, and container growing. 'Stakeless' is another of this type.

'Tomboy,' one of the larger meaty tomatoes.

'Red Oxheart,' 'Pink Ponderosa' (a pink), and 'White Queen' (a white) tomato slices in a relish dish are ornamental as well as delicious.

White Tomatoes

These have a rather eerie quality that many of us never quite get used to. They are low-acid tomatoes of regular, or standard, size, and only a few growers offer seed.

'White Beauty' is one variety you find in catalogs, and 'Snowball' another. They're something of a novelty.

'Roma' is probably the best known of our Italian tomato sauce tomatoes.

Yellow Tomatoes

Most often offered by seedsmen are the small yellow varieties called pear or plum. They are fine for pickling, delicious in salads, and generally mature in about 70 days.

'Yellow Pear,' 70 days, is distinctly pear shaped, and only 1 to 2 inches long. The fruit is borne in clusters. 'Yellow Plum,' 70 days, is similar, but rounder in shape, with 1½-inch fruit. 'Yellow Tiny Tim' is another of these pale, charming miniatures, and is suited to patio and container-growing.

Low-Acid Orange Tomatoes

Of all the varieties, the orange—really, a golden orange, often called yellow—standard-size tomatoes are those from which we expect the lowest acid content. High in Vitamin C, too. 'Orange Queen,' a beefsteak-type tomato that matures in 75 days, and 'Golden Queen,' which matures later, 78 days, and is more on the yellow side, are two types found in many seedsmen's catalogs. 'Jubilee,' 72 days, is a bright golden variety, said to be exceptionally high in vitamins A and C. 'Sunray' is a grower's offering in a disease-resistant orange. Variations on the orange tomato theme include 'Golden Jubilee,' from Nichols Garden Nursery, in Albany, Oregon.

Tomatoes for Canning and Making Tomato Paste

Catalogs usually single out their "paste" tomatoes, as well as those for canning. Paste tomatoes are smallish, pear-shaped, thick-fleshed, and not as juicy as the standards.

Some varieties offered by seedsmen include 'Nova,' a type that matures in 65 days; 'Veeroma' (a takeoff on Roma); 'Roma,' which matures in 75 days and is grown everywhere; and 'San Marzano,' which is slower and a little larger.

Tomato varieties most commonly offered for canning and catsup are 'Heinz' and 'Campbell.' 'Heinz' is disease resistant, as are some strains of 'Campbell,' and matures in about 75 days.

Climbing Tomatoes

Many of the varieties noted as vines will climb trellises. Two kinds recently advertised as meant especially for going up trellises are 'Climbing Trip-L-Crop,' and 'Jung's Giant Climber.'

Disease-resistant Strains

All growers offer many of their best and most popular varieties in strains bred to resist fusarium wilt and verticillium wilt, two diseases common to tomatoes. Pick the variety you like, then look for strains resistant to whatever tomato problems your area may have, if any.

Most growers offer a variety called 'Rutgers' for areas where disease is a tomato problem. 'Veermore,' 'Veeset,' 'Vision,' 'Veebrite,' are the names of disease-resistant varieties offered by R. M. Shumway, Rockford, Illinois, seedsman. These are so important that many other growers group their disease-resistant plants in special categories. 'Campbell' (a canning variety) tagged with the number '1327' and 'Campbell 38,' another canning variety, are also typical of varieties specially bred to offer the public special-purpose types that resist tomato diseases.

'Rutgers,' an old standby. New varieties in which Rutgers is one parent preserve the fine taste and texture and are producing more disease-resistant plants. Below: 'Ultra Girl VFN,' a disease-resistant, early-staking variety.

Ground Cherry and Husk Tomato

These are the names given to a relative of the tomato, which can be grown wherever tomatoes do well. The kind offered by seedsmen generally produce yellow fruit about the size of a cherry, encased loosely in a husk. A novelty, but not a tomato.

Special Varieties for Container-Growing

Whether grown indoors or outdoors, tomatoes are most fun when you get to choose the varieties you are going to nurture. That happens when you start your own seedlings indoors six to eight weeks before planting time.

For hanging-basket plants and very small containers, select varieties such as 'Tiny Tim,' a plant that grows little more than 12 inches tall, and bears cherry-sized fruit. The next size up is probably 'Burpee Pixie Hybrid,' whose fruit grows in clusters that are very decorative, little flecks of red about 1¾ inches around on an 18-inch plant. Another good basket plant is 'Early Salad Hybrid' with fruit a shade smaller than 'Pixie Hybrid' growing on very low plants with a 24-inch spread.

The 'Patio,' a VFN F1 hybrid, is a medium to large size tomato specially bred to container growing or planting in small-area gardens.

Jeannene McDonald picking red plum tomatoes growing in containers alongside miniature yellows and reds.

Among good choices for decorative purposes is 'Patio Hybrid' which produces 2-inch fruit on a compact plant about 30 inches tall. 'Presto Hybrid' is almost as tall, and has fruit a little smaller. Growing on a tidy vine, 'Small Fry' is the tiny tomato to choose if you want to train four plants on a trellis: It covers vines 3 or more feet tall with hundreds of tiny 1-inch fruits that mature in decorative clusters.

For wooden planters and big containers, select any regular-size tomato that suits your purposes. Tomatoes growing outdoors in a garden are generally given three feet all around for growing space, and that, or a little less is what they should have if container-grown; this doesn't mean they need three feet of *container* space all around—just enough container space to comfortably hold the 2 gallons of soil needed to hold the roots and support the plant, plus enough air space and enough staking so they can grow upright in the sunlight.

In selecting tomato varieties for larger containers, beware of varieties that mature giant tomatoes weighing close to one pound. The larger the fruit, the stronger the stake holding the plant upright must be. Fruit of 'Spring Giant,' and 'Fl hybrid,' for instance, can weigh in at three-quarters of a pound each, and the plant needs a stake 6 feet tall and 2 inches square firmly anchored. A stake that size requires a big container to support it, and I do mean big. If you have a sturdy drainpipe handy to train plants of the very large-growing tomatoes on, then go ahead and plant one.

3
How to Handle Seedlings

Starting Seedlings Indoors

You can start tomato seeds in sunny windows or under fluorescent lights. The time to start seedlings for tomatoes to be transplanted to the outdoors is six to seven weeks before the usual date of the last killing frost in your area; six weeks for early, or quick-to-mature, varieties; seven weeks for slow-growers.

You can start the seedlings in peat flats or in old plastic refrigerator trays, in broken cups, or plastic flats left over from last year's garden center purchases. Paper cups or cut-down milk cartons are handy because you can remove the bottoms without disturbing the tender roots. Also, leaving a ring of the container above the garden soil prevents cutworms—which operate near the surface—from mowing down a seedling.

Plant seeds in the potting soil mixture, given on page 28 for basket plants, or on page 33 for general container use. You can start the seedlings in pure vermiculite or sphagnum moss. What growing medium the seeds are rooted in doesn't matter as far as sprouting goes. However, if you want to keep the plants until they reach transplant size in the container in which the seeds were sprouted, you'd best select soil that is dense enough to hold the plants upright as they grow into seedlings. Or else you'll have to transplant seedlings once to a proper supporting potting mixture, and once again to their outdoor containers. Pure vermiculite and sphagnum moss I have found not dense enough to support big seedlings. Using starter cubes of pressed peat is one way of sowing in containers that can be set into garden soil directly. You can also use flats with dividers for sowing seeds so that you do not have to cut intermingled roots apart when it is time to transplant finally into the garden.

With the potting mixture in the container, dampen the soil from the bottom by placing it in a sink of water. Then sprinkle the tomato seeds (they're small) in straight rows if you are planting in flats, or in a circle if you are planting half a dozen seeds in a 6-inch pot (my preference for starting tomatoes). By adding a bit of special powder from the seed store you may prevent "damping off"

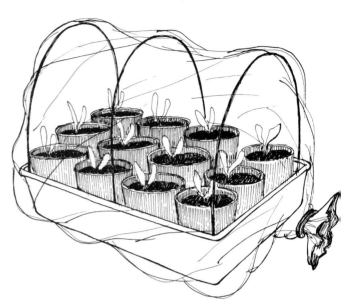

Start seedlings in sphagnum moss, potting soil mixtures, or vermiculite. Space seeds about 2 inches apart. Dampen soil by placing in a sinkful of water, before planting. Cover the seeds with a scattering of potting medium, and then slip into a plastic bag (below) or cover loosely with plastic. Keep the trays at temperatures of between 70 and 75°F.

Seedlings may be started in individual peat pots. A slightly more expensive method, and one that requires more space, it has the advantage that each seed has lots of growing room. Start 2 seeds in each little pot, and as the seedlings develop their second set of leaves, remove the weaker of the two.

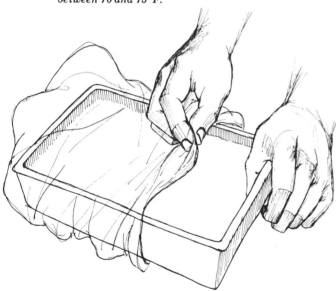

which probably results from too much watering. Damping off kills the seedling as surely as a cutworm.

Cover the seeds with vermiculite, sphagnum moss, or a very light covering of potting soil. Cover the top of the

container with plastic, loosely and place in a warm room—70° to 75°F.—to germinate. They don't need sun at this stage, but appreciate some light. Keep the soil damp, but not soggy.

When the seeds have sprouted little pale stems, and the stems have developed four leaves, it is time to move the containers to a sunny window, and to remove the plastic covers from the containers. Or you can set them under fluorescent grow lights instead, described on page 30, if you have them.

As the seedlings become accustomed to the sunlight, they'll bend toward it. To keep them growing straight, turn the containers every time you water. Continue to keep the soil damp, but not soggy as the plantings grow.

If the plants begin to yellow, allow the soil to dry out, almost to the point where the seedlings wilt, then water with an ammonium nitrate solution at half

Seedlings may also be started in growing cubes. These require no soil or other planting medium: just drop a seed into the cube. After planting, water the cubes thoroughly from the base and place them on a tray in a plastic bag. The seedlings will be ready to transplant when they are about 6 to 8 inches high (right).

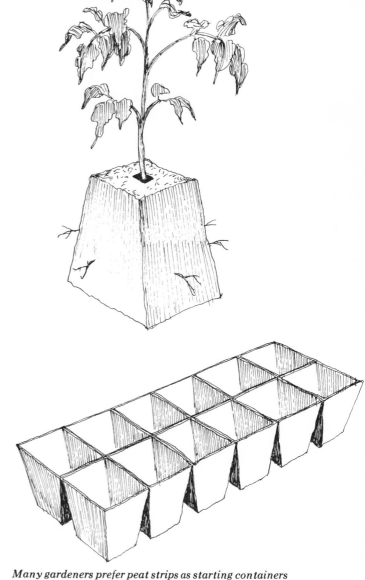

strength. You'll find fertilizers containing ammonium nitrate at your garden center: Read the label, reduce the dose to half, and that will give you an ammonium nitrate solution of the proper strength.

If the seedlings begin to crowd the flat or the pot, it is time to thin. For me this is always a painful process. I never can bear to kill seedlings, and often end up transplanting them with tweezers. I don't recommend it, as you usually end up with more tomato seedlings than the house windows have sun for, and more tomato plants than your garden really needs.

Keep thinning the flats or pots as the plants crowd each other until you are down to the number of plants you plan for transplanting. Then, if the plants are still crowding each other and inhibiting growth, transplant seedlings to separate pots. Think back on flats of tomatoes. Seedlings you've seen in garden centers.

Many gardeners prefer peat strips as starting containers for tomatoes. The cubes keep seedling roots contained so that at transplanting time there's no tearing apart of roots to harm or check plant growth — you just cut, or pull, the cubes apart, and plant.

They were growing in pretty crowded conditions and they were okay, right? Well, you needn't thin more than that.

How many seeds to sprout—Figure out the number of each variety of tomato you want to end up with outdoors, add two to the number and plant three times as many seeds as you want plants. As the plant seedlings develop, thin the flats or pots until you have only the desired number left. It's educational to give your extras to neighborhood children.

Buying Seedlings at the Garden Center

If you get into tomatoes too late to start your own seedlings for this year (six to seven weeks before the date of the last killing frost in your area), select plants from the local garden center, but select with care.

First consideration in buying started tomato seedlings is to make sure the plant you are buying is of a variety that suits your growing possibilities and your purposes. Most if not all 'Beefsteak' tomatoes are less tasty than standard varieties. Italian paste tomatoes also can't compare and are bred for a different purpose. This means that to buy correctly, you should study pages 11 to 16 and your garden catalogs, so you'll know which varieties suit your needs before you select from the garden center offerings.

Generally, garden centers offer a few of the cherry tomato types, a few standards, a few great big tomatoes, and on occasion some exotic types, like the little yellow pears, plum tomatoes, or the meaty Italian tomatoes. You may have to tailor your needs to their offerings. That's easiest if you've figured out before you go shopping which type will work for your garden space.

Once you've settled on the variety, or varieties, you want, look over the plants offered with care. Late in the spring, tomato seedlings are offered by the dozen in flats of peat, plastic, or wood, and also in individual pots. The individually potted tomato seedlings are usually the strongest looking. They haven't been crowded on all sides by brother tomatoes. They also are the most expensive. Those growing in flats may be smaller, and spindlier, but if you get them growing in their permanent garden locations early enough in the season or in a good healthy condition, they'll soon match the plants growing in individual pots. I usually buy one or two individually potted plants of the cherry tomato variety. These have a head start and, because they mature early anyway, produce crops in just weeks, while the standards I buy in flats are still growing up. It's nice to have about six plants of extra cherries planted in the garden and invite children and guests to sample the ripe ones in order to keep your other tomatoes safe from temptation.

Select sturdy, stocky-looking plants. Avoid spindly, leggy plants and flats showing a lot of yellowing leaves at the base of the plants. Plants that have lots of little yellow flower buds will produce tomatoes sooner than plants of a similar variety that haven't yet developed flower buds. The tomatoes develop from the base of these buds once they have flowered.

Often enough you can even find seedlings on which tiny green tomatoes are already beginning to swell. These will produce fruit sooner than plants just beginning to flower, so let them be your preference.

Transplanting Seedlings

Transplanting seedlings, whether to the open garden or to containers, isn't difficult. Tomatoes don't mind transplanting, and, in fact, seem to thrive on the change of soil. Probably because they may have begun to outgrow nutrients available in the potting soils they first grew in.

If your seedlings are growing in flats or pots indoors, harden them off before transplanting them to the outdoors. If they are going outdoors in new containers, and will remain container plants, you can transplant them to their permanent containers any time you think they've outgrown their starting flats. If they are going from an indoor growing flat or pot to the open garden, then one week before transplanting time, harden them off. Hardening off is a simple process: Just place the flats outdoors in a semishaded corner where there isn't too much wind for a day or two, then move them to a sunnier, more exposed location for the rest of the week. If very cold weather threatens the first day or two of the hardening-off period, bring the plants to the shelter of the porch or even indoors. Sudden cold will check their growth.

Many growers transplant tomato seedlings three times: once from starting flat to 3-inch peat pots; from peat pots to large clay or plastic pots; and finally to the open garden.

Whether the transplanting procedure is for one of these moves, or directly from the flat to the permanent container or the open garden, the moves are the same. Soil should be *slightly* damp, not wet, not bone dry. If the plant is growing in a flat, use a sharp knife to cut out a square containing the roots around each seedling. This will cause less harm than will tearing the roots apart, however gently and lovingly you tear. Lift each square out, retaining as much soil as possible, before planting. If the plant is growing alone in a pot, place your hand across the top of the pot, with the center stem between your fingers; turn the pot upside down; give the rim a sharp knock on a sink corner or a wooden board; and the soil and roots will come out in a compact lump. Don't unpot the seedlings, however, until you are ready to plant.

The hole for the tomato seedling should be slightly deeper than the root ball, about an inch or two, depending on the size of the plant. If you are planting in a container, fill the bottom of the container with 2 inches of gravel, then add enough potting soil to bring the tomato root ball to a level 2 or 3 inches below the rim of the container. Place the root ball on the soil, fill the container half full with soil,

and add a transplanting formula, such as Transplantone, to water at tepid temperature. Fill the hole with the water, and allow it to drain through. Dribble enough additional soil around the edges of the root ball to fill the container, water once again with transplanting liquid, and allow the plant to rest for a day or so. When the soil has dried a little, firm the root ball into place with your hands.

If you are transplanting to the open garden, dig the hole 1 to 2 inches deeper than the plant root ball, and set the seedling upright in the hole. If the seedling is leggy (branches begin very high up the stem), set the root ball in the planting hole at a 45° angle and bend the main stem upward, so that branching begins just above the soil level. New roots will form along the stem where it was bare before. Fill the hole with tepid water, to which you have added a transplant formula. Allow the water to drain out. Fill the hole with soil to within 1 inch of the garden soil level, add transplanting liquid, and when it has drained out, gently firm the tomato root ball into place.

Water often the first week or two after transplanting, but not enough to keep the soil soggy. This is especially true for containerized plants.

When to Plant Seedlings

In most of the upper South and the North, tomatoes are planted in mid-spring—as soon as the ground has warmed—and will keep growing until early fall and the arrival of the first hard frosts.

What "early spring" means depends on where you live.

In areas that are frost-free most of the year, like southern Florida and some microclimates along the Gulf of Mexico, tomato seedlings can be set out in January, February, and March. Second crops can be set out again in July, August, September, through December.

In the belt just north of this, areas where the growing season is 260 days, and the last frost is about February 20, set tomatoes out in February or March and try for a second crop, if you like, planted in August.

Where the last hard frost is about March 20, set seedlings out from late March through early May. You can try for a second crop, planted as seedlings in July.

For the rest of the country, there is time only for one crop annually, set out from April through June. This is the belt that includes the southern portion of New England. Here the last hard frost is about March 25. Northward, where the last hard frost is about April 10, May is the planting date for tomatoes.

If you're a little uncertain of the weather, and want to get the seedlings in anyway, plant them and cover them with big brown paper supermarket bags. Tear the bag tops to let in a little air for the tomatoes to become acclimatized, and after a few days remove the bags completely.

Head Start Program for Outdoor Tomatoes

It's hard to say why competitiveness rears its ugly head in the growing of tomatoes, but you'll find it does. Once you get involved with wanting ripe tomatoes *soon,* you start wanting ripe tomatoes *sooner* than anybody else, or at least, sooner than last year. Most tomato enthusiasts have all sorts of ways to run this race. Here are some of those I've tried:

Planting mature seedlings: Picking the largest possible started tomato plant at the garden center, one loaded with yellow blooms, green buds, and lots of baby tomatoes, is one way to get a crop soon. Choose early varieties 'Pixie' and 'Small Fry,' tiny tomatoes that are ready in about 52 days from seed. Planted as seedlings, you'll have tomatoes in 45

days, depending on the warmth of the season and the sunlight available. Greenhouse nurseries have expensive pots with fairly advanced fruit on 2 footers. And if you have a small greenhouse or winter pit, you can really beat your neighbor if the planting area has southern exposure protected from wind and is not in a low cold-gathering spot but, preferably, on a hillside.

Fertilize well: The quickest tomato crop I ever got came in just weeks—and they weren't miniatures, either—they were the result of successful fishing. One autumn we had filetted and frozen a lot of flounder caught at the mouth of a little river in Westport, Connecticut, on Long Island Sound. The fish cleanings, fish heads, bones, etc., we planted in a row in the garden. In the spring, as soon as the weather had warmed, we planted the tomato seedlings there—and they really grew like crazy and produced heavily all season long. The Indians planted a fish with corn, beans, and squash (the last two together so their vines grew up the corn stalk).

Defy the planting season: Normally, tomatoes can be set out only after the ground has thoroughly warmed and all danger of frost is passed. However, there are ways to successfully defy this rule. You can grow tomatoes in a greenhouse, if you have one handy. You can grow them in hot beds, small wood-framed and glass-topped enclosed garden spaces heated by coils carrying hot water or electricity or by an underlying layer of fresh manure.

You can grow them in cold frames, which are similar to hot beds, but are not heated.

Hot beds and cold frames aren't normally of a height to accommodate tomato plants as they mature, but you can custom-build one to create an individual hothouse for your tomato seedlings. We did it one year with storm windows. We set the windows into the garden soil in positions that created a square of glass around each plant, and topped the frames with heavy-duty plastic which could be lifted during good weather days to let excess heat and moisture escape. You can take head-start programming a step farther, and in the fall, prepare the planting hole for the tomato seedlings, and in early spring set the seedling out. If you can find fresh horse manure, which heats up as it decays, incorporate it into the soil in the bottom of the planting hole, in a 3-inch layer. Top with 6-inch layer of soil. Plant the tomato root ball, fill the hole with soil, and create a storm window or plastic house, ventilated during the heat of day.

Despite all of this sound advice, I'd like to say that all else being right, it is my experience that generally tomato plants grow and produce quickly only after the weather has become tomato weather.

Basket-grown tomatoes produce abundantly and are ornamental plants to hang in sunny windows.

4
Special Ways to Grow Them

Tomatoes Grown in Containers

Take heart, city dweller and you who dwell in gardenless places, you, too, can grow your own tomatoes. Not bushels and bushels and bushels, perhaps, but enough for summer salads, and happy-hour snacks, and to toss into the meat sauce as it bubbles to completion.

Actually, if you want to devote *every* possible space in and around your home, be it high-rise apartment or urban postage stamp, to the care and nurture of tomatoes, you *can* have them by the bushel.

In hanging baskets, you can grow cascades of little tomatoes in every sunny window, from every patio or porch overhang. You can even hang them from sturdy hooks outside the windows and let them grow down the sides of the building. (If you are willing to risk your neck to harvest the crop from the bottom of the vine, that is.)

For basket-grown tomatoes, all you need is air and sunlight. No real garden space.

Under fluorescent lights, you can grow the small varieties of tomatoes indoors, literally anywhere you can find space to stand the container and the lighting equipment. You can grow them as room dividers, as front hall ornamentals, in the linen closet, your dressing room. In the basement or the attic or by the elevator (if the super will let you).

For indoor tomatoes, all you need is containers, fluorescent lights, and ventilation.

The bigger varieties of tomatoes, you can grow anywhere outdoors as long as there's lots of sunlight and a spot big enough to stand the container on. The container can be anything that holds at least 2 gallons of potting mixture: raised wooden planters, wooden boxes, tubs, rough wooden flats, crates from the grocery store, gorgeous cement urns, tailored redwood boxes. The spot can be anywhere on the patio, the terrace, the porch (sunny), the fire

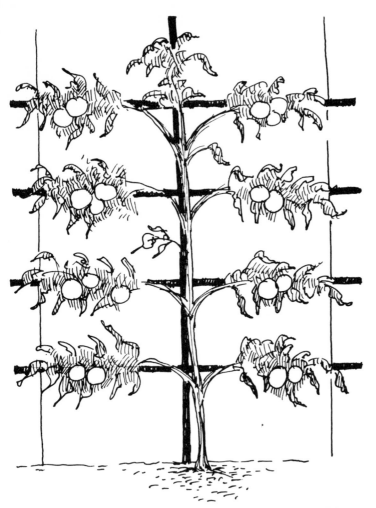

Tomato plants can be espaliered in a horizontal-T form, whether grown in a container or in the ground. To make an espaliered tomato, you must create staking that conforms to the espalier shape you have chosen, and train the branches as they develop to the stake form. Tie branches loosely to the framework with soft green wool or raffia, and keep all suckers severely pruned at least once a week.

places, providing there is enough light. Tomato plants require at least 6 hours or more of direct sunlight daily. Those growing in bright sunny places develop sturdy, well-branched shoots which bear lots of fruit. You can tell if yours don't have enough light by the fact that they'll develop long, poorly branched shoots with few flowers and therefore few fruits.

Basic Equipment for Potted Tomatoes

For the growing of tomatoes in containers, you need potting mixture (that's the same you buy to grow houseplants in), tomato seeds or started seedlings from the garden center or the local plant shop, and knowledge of how to water down fertilizer. The container, in addition to holding at least 2 gallons (8 quarts—32 cups) of potting mixture, should have holes for drainage or be made of a material you can drill or puncture for that purpose. You make up for the lack of ample soil by lots of watering and by including lots of fertilizer in the potting mix.

For smaller plants of the smaller varieties you can use 1-gallon plastic bottles or clay pots or bottles with the tops cut off. If your tomatoes are to cascade downward as they do as basket plants, you need nothing more than supports for the basket and a hook to hang it from. If nothing more charming is available, you can make a tomato garden in one of those big, heavy-duty garbage bags, near an existing support or a trellis, for instance. If the tomatoes are to climb upward, as they must in most containers, then you need some form of stake or support.

Growing Tomatoes in Hanging Baskets

Don't laugh, they're pretty. And if you start your tomato baskets a few days

escape, the roof. Containerized tomatoes should be staked or you can set them to climbing, with a little help, up the trellises, over arbors, gazebos, along fences, or tie them to lampposts, mailboxes, children's swing supports, the penthouse air ducts. They can be espaliered, too, tied and pruned into those charming, stylized shapes you see in old-fashioned continental gardens.

While all this may sound farfetched, it isn't. You can grow tomatoes in all these

Tomato trained to a fan shape, either in a container or growing in the ground.

Small tomatoes of a low-growing bush variety make a handsome display grown in a big planter equipped with stakes tied tepee fashion. There's one stake for each tomato seedling, and cross-stakes to which lateral branches may be trained. An ornamental way to handle tomatoes growing in a small garden or on a terrace.

Containerized tomato staked to an ornamental trellis. An advantage to containerizing your tomatoes is that you can hide the container in a sunny corner until it has developed enough fruit to be considered decorative, and then move it to a prominent spot on the patio or terrace.

A container that will hold 32 cups — 8 quarts — of potting soil, or tomato soil, is big enough to grow any of the smaller varieties of tomatoes. Since the tomato needs staking, except for the smallest varieties such as 'Tiny Tim,' the container must be big enough and deep enough to support the vine as it begins to bear fruit. However, you can grow larger tomato plants in small containers, containers too small for a suitable stake, if you place the container near a strong support to which the developing vine can be attached.

Step 1 in planting a hanging basket tomato is to line the container with broken shards of clay pots to keep soil from flowing through drainage holes, and to provide drainage for the plant roots. Cover shards with potting mix.

Firm the tomato seedlings into the potting soil. Note that these seedlings are being transplanted just after they have developed the second true pair of leaves. Growing tomatoes in baskets as shown here is an experimental process introduced by U.S.D.A.

apart, you'll have crops over a long period.

To create a tomato basket you can go out and buy plastic baskets for hanging plants, like those shown in the photos on this page. Or, you can use any container that will hold at least 2 quarts of soil. Plastic bottles or milk containers that will hold a gallon will do. Use a sharp kitchen knife or scissors to remove the plastic bottle tops, and punch four or five holes in the bottom to provide drainage. Tie a cord or hemp bag—or a macramé holder, which you can buy or make—around the improvised container, and poke three sturdy wires (each about 30 inches long) into the holder. Twist the wires together at the top 3 inches, and bend the wires over to form a handle. Now you can hang the handle on a sturdy hook, and your hanging basket is hanging. (I cover my plastic container with bamboo place mats held on with a rubber band—or use anything you fancy.)

The potting soil described on page 33 is suitable for hanging-basket tomatoes, but you can make a lighter version for this tomato project by blending equal parts peat moss and vermiculite. The fertilizer recommended by the U.S. Department of Agriculture, which experimented extensively with growing tomatoes as basket plants, is a mixture of ½ cup of pulverized dolmitic limestone, 4½ cups of 20 percent superphosphate, and ¼ cup of 5-10-5 fertilizer, for each bushel of potting soil.

With your potting soil at ready, cover the drainage holes of the basket with 2 inches of bits of broken clay pots, or shredded plastic pebbles. Fill the container with moistened soil mixture to within 1 inch of the rim. Now the basket is ready for your tomato seedlings.

If your seedlings were started in individual containers, put your fingers across the top of the soil on either side of the tomato stem, turn the container upside down, and knock the rim sharply against the corner of the table or sink to loosen the soil. The root ball will now slide out of its pot. If your seedlings are growing in a flat, use a sharp knife to cut a square of soil around the root system of each. Don't tear them apart. Loosen the outside network of roots gently with a fork around the soil ball before planting. Scoop a hole in the basket soil deep enough and wide enough to hold the soil ball of the seedling and firm the soil ball into the basket soil with your hands. Press hard. Hang in sunny window.

The U.S.D.A. method for growing basket tomatoes includes a deliberate program of wilting. It's a view of tomato-growing that I've not experimented with, but which may interest you.

The U.S.D.A.'s purpose in allowing the tomato plants to wilt a little is to help new seedlings adapt to their basket life. The method is to water the plant as soon as it is in its basket, with a solution containing one-fifth the fertilizer recommended for houseplants. They recommend waiting to water again until the leaf color changes from dark to pale green and the plant begins to wilt. Each time you water—probably every 3 to 7 days—wait until the plant wilts, and use the same amount of water and fertilizer. The treatment is designed to help basket-grown tomatoes toughen, and become better able to withstand sudden changes in environment: hot, dry winds, weekends when the grower vanishes on holiday.

Placing a basket plant is simple: It must have 6 hours of direct sunlight daily. Or more. You can locate it in a southern window, under a porch overhang, or from the children's swing poles on a high-rise roof. As the plant develops, remove some of the branches to prevent crowding and to promote development of new flowering shoots. (See page 39.) Turn the container, if it is in a window or under a porch overhang, at weekly intervals so that it will develop symmetrically and balance out the lopsided effects of uneven sunlight.

You should have ripening fruit any-

U.S.D.A. basket tomato experiment includes watering the plant only after it shows signs of wilting, and watering with a solution containing one-fifth the fertilizer recommended for houseplants.

H. Marc Cathey, horticulturist with the U.S.D.A., is shown in Beltsville, Maryland, greenhouse with a collection of tomatoes growing in hanging baskets.

where from 45 to 60 days after you start the plant, if the season is warm and there's sunlight.

Growing Tomatoes under Fluorescent Lights

You can start, grow, and ripen tomatoes under lights—which means, you really can grow a crop indoors. Among the varieties that will mature fruit under such conditions are the little ones like 'Tiny Tim.' A whole tray full of these miniature ripening fruit is a pretty sight, just as pretty as Jerusalem Cherry and pepper plants, both popular houseplants. However, the tomatoes won't go on indefinitely bearing fruit, so the indoor to-

A two-tiered fluorescent light stand containing a gardenful of seedlings in different stages of maturity. Here, ready to set out, are tomatoes, cabbage, broccoli, and several types of lettuce, as well as flowers in individual pots.

mato garden can't be counted on as a permanent display.

To grow tomatoes indoors you need a fluorescent lighting fixture of the type sold commercially under names like Gro-Light. The basic setup is a standard industrial preheat fixture with two 48-inch, 40-watt tubes. One tube should be daylight and the other natural white. Or one tube can be daylight, with one plant growth lamp such as Gro-Lux. You also need a reflector suspended about 18 inches above the surface of the bench or table on which the tomatoes are to grow. Keep the lights on for 16 hours out of every 24 for tomatoes. They'll grow well and compactly.

A pair of 40-watt fluorescents, 48 inches long, will effectively light a tomato garden 2 x 4 feet. You can buy a timer that will switch the lights on and off to create a regular 16-hour day. Or you can do the switching manually. If you are going to be absent for any number of days during tomato-growing season, the plants will suffer from lack of light, if you leave them at off. They will also suffer from too much light exposure if you leave the fluorescent at on.

If possible, buy a reflector that reaches outward 15 inches from the fluorescents. Narrower reflectors are not as effective in focusing light onto the plants. You can create a reflector, by backing the lights with a piece of plywood extending outward 15 inches on either side. Coat this with a flat-white, rubber-base paint that diffuses light well.

If you are planning to install your fluorescent in a dark place, a closet, a basement, or a dim hallway, the plants will respond best if the surfaces around are painted white to improve light reflection. Mirrors in the background and at the ends of growing sections also reflect light.

To keep the lights at optimum efficiency, wipe the tubes down weekly with a damp cloth; dust can dim them considerably.

Humidity is important to an indoor tomato garden. At temperatures above 68°F., air dries out the plants. Offset this by placing the potted tomatoes on a tray of pebbles filled with water. You can make the tray attractive by covering it with florists' moss. Another solution to lack of humidity indoors is to set glasses of water among the pots. Misting the plants as you do houseplants once or twice a week is a great help.

Tomato seedlings started under fluorescents can be transferred to the natural light of a window, or outdoors, without ill effects. Move the plants to light and moisture conditions approximating those where they were born. Don't for instance, set plants fresh from fluorescents under the blazing full sun of a hot summer noon. Move them first to a slightly sheltered area, and then to their permanent sunny home. The 16-hour day your fluorescents provide for the plants are meant to approximate the light they would receive if they were outdoors that long, in sunlight.

To get fruit from the plants, you'll have to imitate another condition that occurs naturally outdoors: movement of the plants' branches. Tomato blossoms require pollination from other tomato blossoms on the same plant before they can set fruit. Outdoors, the breezes shake the blossoms and scatter pollen over the plant. And insects help tomato pollination, as well. Indoors, with neither breezes nor insects, you get the same effect by shaking the plants gently a few times a day when they are in bloom. Or else, try spraying the blossoms with a fruit-setting hormone, which most garden catalogs and centers offer.

You'll also have to approximate the growing weather for tomatoes in temperature: Tomatoes flourish in temperatures above 70° F. This means they'll succeed indoors at that temperature, or even higher, but they will sulk if you try to grow them in a cool basement. Many of the vegetables we start under fluorescents indoors prefer the cool of basements, but not tomatoes.

5
The Outdoor Garden—
What to Do

Soils for Tomatoes

Whether you are growing your tomatoes in a potting mixture in containers, or outdoors in the open garden, make the effort to create a soil especially for them. Or buy one.

Because tomatoes are by far the most popular vegetable for home gardens, garden centers now display both fertilizers and soils especially prepared for tomatoes. One soilless mix recently introduced, called Tomato Soil, is completely supplied with all the nutrients necessary to keep the plants producing through the season. It needs no further fertilizing. You can use this soil for containerized plants. Or you can create a planting pocket in the open garden and fill it with tomato soil.

Garden soils suited to tomato crops are those with a pH of between 5.5 and 6.5. That's a broad range. At 5.5, the soil is quite acid. At 6.5, the soil is heading toward neutral, pH 7.0. Which is to say that your soil probably is okay. You can check it with one of the little kits that give pH readings, or you can have it checked out by the local Agricultural Extension Service. In most states they are located at the state university, but you might find an agricultural service agent in your own town. Take samples of the soil to him.

Local soil is not likely to be okay for tomatoes in desert and semidesert regions where rainfall is very low. In such areas, various alkaline minerals remain in the upper layers of the soil, increasing year by year. While many plants will tolerate moderate alkalinity (delphiniums and lilacs thrive on it) only a few grow really well in soil close to the neutral point.

Another instance in which the local soil may not be okay is if it has recently been improved with many applications of lime and bone meal and other fertilizers that are alkaline.

Local soil is likely to be good for tomatoes if it has been improved with peat moss, compost, wood chips, all of which tend to incline the soil toward acidity.

The easiest way to improve garden soil for tomatoes to be grown in the open garden is to dig ⅓ sharp sand and ⅓ humus, along with suitable fertilizers (see pages 34–35) into planting areas large enough to contain the plant roots at maturity—about 18 inches around and 18 inches deep. Or, you can dig a hole that size and fill it with specially prepared potting soil that you've purchased—or mixed yourself.

To make your own potting soil for tomatoes, combine ⅓ garden soil, or non-alkaline commercial potting soil, with ⅓ decomposed organic matter, such as peat moss, or compost, and ⅓ part sharp sand.

The humus bits act as a sponge to retain moisture and they keep the soil from drying out. Tomatoes require lots of humus, whether growing in the open garden or in containers. But especially when they are growing in containers.

"Sharp" sand, if you look at the grains under a magnifying glass, has jagged edges, and these make air spaces between the grains, which allow roots to breathe and water to circulate. You can use vermiculite instead of sharp sand if you want to avoid excessive weight in containerized plants. Vermiculite is indicated for basket tomatoes, as you will see on page 28, where special potting combinations for basket tomatoes are described. However, precisely because vermiculite lightens the soil, it creates a mixture that offers less support to the tomato plants. If you are growing tall, heavy plants in windy places, sand is a better choice.

The Water They Need

Since the fruit of the tomato vine includes lots of water, the plant producing the fruit obviously needs lots of water.

Watering the tomato seedlings at transplanting time is described on page 21.

Watering for maintenance purposes through the season isn't arduous, but it must be conscientious, or baby fruits getting ready to mature will find their growth checked and slowed.

Any food plant in the open garden requires a really good watering once a week. If the sky does it, you needn't. If the sky doesn't, then you must.

A "good watering" can be defined as one that gets moisture down at least 8 to 10 inches into the ground. Sometime when your garden is bone dry, stand and water your garden for 20 minutes with a garden hose. Then poke your finger, or trowel, down into the soil to see how far the moisture has penetrated. You'll be surprised that the dampened soil may be less than an inch deep. It takes a long time to get a very dry garden soil humid right down to where the plant roots are growing.

To check exactly how long it takes to water your garden to 8 to 10 inches, run your automatic garden spray for an hour, and set under it a peanut butter jar at a point where the full fall of water is effective. Some peanut butter jars have inches marked on the sides. After an hour, check the jar to see how many inches of water have fallen. Between 1 and 1½ inches of water will furnish moisture down the 6 to 8 inches necessary for most vegetables. For tomato plants, which are deeper-growing than small vegetables, a 2 to 2½ inch rainfall is best. Find out how long it takes your watering system to put down 2 to 2½ inches of water before you deliver your tomatoes into its hands.

Watering containerized plants is surer but more time-consuming; surer, because you do it more often, and because there is less soil to get wet; more time-consuming because it must be done more often. A good rainfall once a week will probably hold a big tomato plant growing in a big container two or three days. In a huge container, it might dampen the plant for five or six days. But a small contain-

The multi-plant method of Leopold Klein, with movable container. Mr. Klein's concentrated fertilizing produces heavy-bearing plants in sixty days. Below, Mr. Klein's container, with plants started indoors, is moved out onto the balcony. These plants are four weeks younger than those above.

er will lose its moisture much more quickly than that.

There are several reasons why soils in containers dry out quickly. One of them is the plant itself: Rootlets are constantly drawing moisture from the soil in the container; porous containers, clay pots, wooden tubs or flats, cement urns, all draw moisture from the soil inside and on their surfaces it evaporates in the sun and the wind.

The only way to be sure your potted tomatoes aren't running dry is to check the soil daily, or every second day. If the surface is dry (as opposed to damp), water your small containers. If the soil is dry 2 inches down in big containers, water again.

But don't overwater. Containerized tomatoes have a lot in common with houseplants, and really are to be treated the same way. More people kill houseplants by overwatering than by any other means. Overwatered soil is soil that is constantly soggy; really wet soil has all its air spaces filled with water, which means the roots will soon be rotted:

Fertilizing Rites

Because tomatoes grow rapidly and produce big crops over a long period on small vines, the soil must be fertile. Nitrogen encourages leaf growth. Phosphates encourage flowering and fruiting. Since tomatoes taste better than leaves, avoid fertilizers high in nitrogen, and select instead fertilizers high in superphosphate. Package labels tell which is which.

There are special fertilizers on the market today composed especially for tomatoes. Incorporate these into tomato potting soils, or into the holes planned for tomato seedlings in the open garden.

If you want to create your own fertilizer, for each bushel of potting mix or garden soil, add: 8 level tablespoons of superphosphate, 8 level tablespoons of cottonseed meal, 4 level tablespoons of sulphate of potash. Mix the fertilizer into the soil a week or two before tomato planting time if possible, to give it a chance to begin to break down and become part of the growing medium. But don't add it too early or the fertilizer may leach out.

If yours is a first-year garden or if your soil is not yet built up with humus, you can still get off to a fast start. Buy 4 yards of well-fertilized farm soil from a friendly farmer. Ask him to add 1 yard of rotted cow manure. Have him dump the truckload (at the cost of building sand) in a sunny spot. Form the soil into mounds 6 to 8 inches high and 4 feet long. With additional lime and any other nutrients advised by your county agricultural agent, you should be able to raise well over a ton of vegetables.

Once your seedlings are growing vigorously, they don't have to be fertilized again. However, you'll get faster fruiting if you incorporate a tablespoon of the fertilizer described above, or tomato fertilizer, or a complete fertilizer high in phosphates, into the soil around the plants once they have doubled transplanting size.

You'll do exceptionally well with potted plants if you fertilize every two weeks. Dissolve the fertilizer into the watering can so that it will be immediately accessible to the plant.

Staking Your Tomatoes

Left to their own devices, standard tomato plants grow upward, from 20 to 30 inches. Then, as leaf growth and branching make them top-heavy, they lean over toward the ground unless staked. Leaning doesn't hurt the plants—actually, the thickness of subsidiary branches helps keep the plants from lying flat on the ground. However, plants that aren't staked seem to mature fewer fruits, probably because the earthbound portions of the plants receive too little sunlight. And, too, the fruit matures less quickly on stakeless plants because they are shaded from the sun by so much greenery. After ripening, fruits on unstaked plants are often overlooked in the thick mat of leaves. With less air movement around the flower blossoms, so sheltered by their tent of leaves, I suspect that fewer blossoms are pollinated and, therefore, fewer fruits are set. That's why unstaked tomatoes are often grown on a low platform of sticks or slats.

Then, if the suckers are not pinched off (see illustration on page 39), many new flowering branches will grow and a heavier crop of tomatoes will result. But the strength of the vine will go into the increase in fruit and the tomatoes will mature later. Unsnipped suckers will give you lots of green tomatoes at the end of

Bare branches piled beside a tomato row provide support and keep fruit from touching the soil.

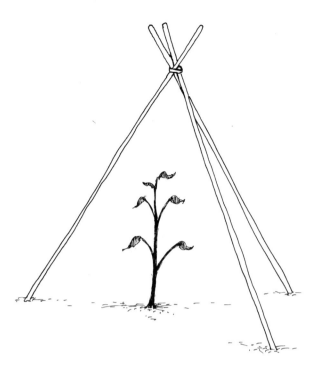

Three-pole tepee stake is often used to support tomato vines of larger varieties. The lead branch is tied at first to a pole, then to the tepee top as it reaches full height. The lateral branches are tied to the tepee poles.

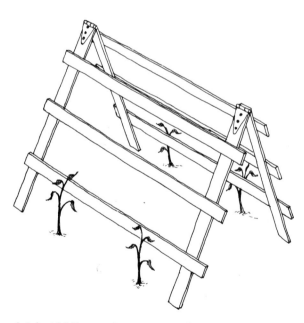

A 6-foot folding sawhorse supports four tomato plants. Well-built, the sawhorse will last for many seasons, is easy to store in winter, and provides excellent support.

the season just before frost. It's a rule of thumb that if you are not planning to stake your tomatoes, you should plant more vines. For all these reasons, most home growers stake their tomatoes.

The ways of staking tomatoes are innumerable. A few of the best known are illustrated on the next pages. The most important point to keep in mind in choosing a staking method is that the stake be thick enough to hold the vines once they have matured and are setting lots of fruit. Plants of the beefsteak type of tomato, which matures big, heavy fruit, must be tied loosely to stakes at least 2 inches square, and set into the ground deeply enough to stand firm whatever happens.

When to stake? I've always found it easiest to transplant and set the stake or stakes at the same time. For one thing, the plant root system is small at that time, and you know where it begins and ends. Later, the roots may have grown outward, and in setting the stake into the ground close to the plant, you may break important roots. Besides, it's just easier on the mind to do both at once.

When tying, leave 6 inches soil space between plant stem and the stake or stakes. When to tie the plant to the stake? When it can conveniently be done. You'll have to tie portions of the plant to the staking system several times during the early growing season, as the plant grows taller. Set the first tie as shown in the illustration on page 37—loosely, in loops, using soft rag strips or bulky wool. Don't clamp the plant tightly to the stake with a wire twist; it needs room to grow. As the plant grows taller, branches thicken, reach outward, and begin to show their future burden of fruit. Tie the harvest branches to stakes at strategic intervals. The hardware and garden centers sell bundles of pencil-thin bamboo stakes dyed green. These stakes won't hold up any vine larger than a miniature,

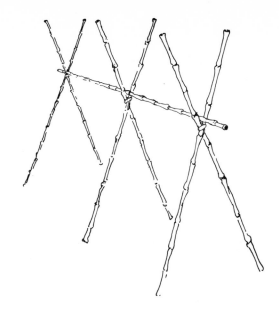

Picket fence makes an excellent support for growing tomato vines. As they grow, the sucker branches can grow
along the horizontal braces. The fence needs to be 6 feet
tall for larger varieties.

Row of bamboo poles, tied at the tops, with a bamboo
cross stake set in the crotches of the ties, makes one of
the most attractive of staking systems for a row of tomatoes.

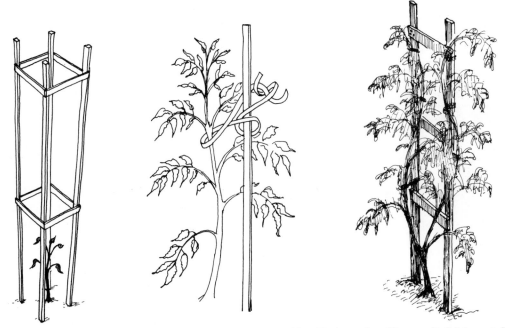

Left: A strong four-pole support for a vine of the heavi-
est varieties. A similar stake can be made by rolling con-
crete-reinforcing wire cylinder and burying it 3 to 4
inches deep, anchoring it along one side with a wooden
stake driven firmly into the ground. Middle: Tie young
tomato seedling to staking with a double loop as shown

here. Use old rags, soft raffia, or soft, thick wool. Keep
the tie loose enough so that tie doesn't cut into or break
the branch, but firm enough to train the tomato into po-
sition on the stake. Right: Two-pole stake for tomato
vine, with cross supports to keep stakes rigid. This is
strong enough for heavy-producing or giant varieties.

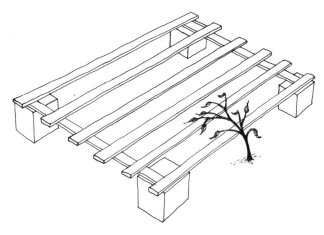

Tomatoes can be grown horizontally on a flat trellis supported by clocks or bricks, or by Y-shaped stakes with slender tree-branch poles laid across them. The suckers may be allowed to grow freely on these unstaked tomatoes and as many as three or four plants can be grown on a single support.

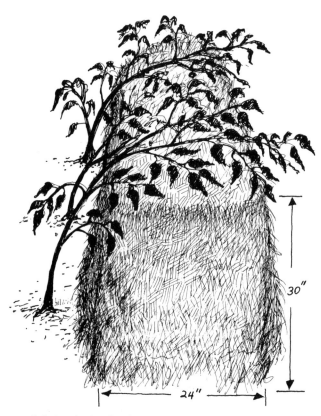

Salt hay is the classic mulch for tomato vines that are not staked. One bale spread 6 to 8 inches deep on the soil will accommodate vines planted 3 feet apart and allowed to "grow wild." The vines will spread over the hay, which will protect fruit from contact with the soil.

but they are handy to support branches that can't be tied to the main stake without crowding the leaves. You can stick them in the ground well beyond the root system.

Mulching to Keep Moisture In, Weeds Out

A mulch is a blanket, an organic blanket such as hay usually placed around the soil at the base of a plant to keep weeds smothered, and the soil protected from the drying effects of sun and wind. Mulches also stop soil erosion in hilly places.

Organic simply means derived from plants or animals. In the context in which we use it today, it generally refers to natural materials that once had life and have decomposed into humus. Peat moss is the most commonly used organic blanket in home gardens, while in truck gardens, black plastic mulch (which isn't, of course, organic, but is effective) is found to be easier to handle when covering big spaces.

Organic mulches have the advantage that they decompose on the side that's in contact with the earth, and so improve the soil as they protect it and the plant roots growing in it. They draw worms, which aerate and lighten the soil.

Grass clippings are sometimes used to mulch around small planted spaces. It's surprisingly hard to get enough grass clippings unless you have an estate—to cover garden rows. For one thing, it takes grass many inches deep—five or six in my experience, to really keep the weeds down. For another, grass clippings often have weed seeds in them, and these turn up next season in full glory right in the soil the clippings were meant to keep weed free.

Buckwheat hulls are an attractive mulch, since their color blends well with soil colors. Seaweed, gathered in quantity on the shore during fall storms, is an excellent mulch. Place it in the garden rows as it is gathered, and plant between these mulched rows the next spring.

If mulching draws moles from nearby woods, plant garlic in the runways to repel them. But if this is not successful, you must remove the mulch and soak the runs with anti-mole chemicals dissolved in water.

Mulches for containerized plants are somewhat less complex to deal with. The soil spaces to be covered are very manageable, and since weeds aren't much of a problem, you only need enough mulch, an inch or two, to keep the soil surface from coming into direct contact with drying air. Handsome stones are often used in a handsome pot.

To Pinch or Not

Pinching is a method of pruning away the "suckers" so that the strength of the plants can be devoted to main mature branches that produce fruit, and to the maturing of the plant itself. Between the main stem and the leaves growing along it—just where they join—there are always extra shoots that will, if allowed to grow, become fruit-bearing and take away strength from the main branches of the main stem. You'll get more tomatoes if you let them grow, but the whole plant will mature more slowly. That's your choice to make, but don't pinch off main branches!

My father pinches out the tomato suckers and his fruit seems to mature more quickly. I don't pinch mine, largely because I never feel I have the time, but mine mature fruit more slowly. Ergo, pinch if you have time, I always say, and get earlier ripe tomatoes—but fewer.

A Stitch in Time

Tomatoes, like people, have health problems: little pests that eat them or lay eggs on them or transmit diseases

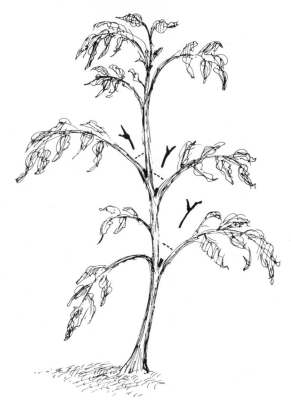

The illustration shows where suckers grow in crotches of tomato branches. The Y shapes are the detached suckers which would become bearing branches if you didn't snip them off. Removal of all suckers once a week helps the plant to ripen fruit early, on the main vine. Prune by hand. Hold the shoot with your thumb and forefinger, twist it sharply to one side till it snaps, then pull it off in the opposite direction.

that harm growth, halt fruiting, or spoil the fruit. However, tomatoes aren't too vulnerable, except in limited areas of the country. Ways to discourage pests and diseases are discussed on pages 42–44.

But before things get to the point where you have to call in the troops, there are some good garden practices that should keep problems from your door.

Like people, plants need air. Damp, soggy, windless corners encourage fungus and other diseases. Airy locations (not windy, just airy) discourage both pests and diseases: pests because they get blown away by the breezes; diseases because the conditions that foster their growth are destroyed by sunlight and air.

Avoid creating excessively humid conditions. This means, as stated earlier, don't keep containers or soil around plants soaking wet. Soggy soil drowns the plant. Soil constantly very wet encourages growth of pests and diseases, especially in hot airless weather.

Hybrid 'Rushmore.' Note the healthy disease-resistant leaves.

Don't spread problems. If tomato leaves, or other garden crops are diseased, or suspected of disease, remove and burn. And avoid touching healthy plants with tools or hands or gloves that have come into contact with diseased or pest-ridden materials. I'm told one should never touch wet tomato leaves. After rain or watering, stay out of the garden. You may spread leaf disease. Separate containerized problem plants, or plants suspected of problems, from the healthy plants and keep them in quarantine until cured. Or else, discard them. At all costs, don't preserve a diseased plant—you will probably regret it. But don't decide a plant must be discarded until you've determined its problem.

6
Making Sure Things Won't Go Wrong

You can have glorious tomato vines, and few tomatoes if any of the following conditions exist:

Setting tomatoes out too early: tomatoes set fruit within a narrow temperature range. Sunny temperatures below 60°F. discourage fruit set. The blossoms don't become fruit, or they drop off. That's why getting a tomato into the ground early doesn't always mean you'll get fruit early. Early varieties will set fruit at colder temperatures, of course. You may be able to foil nature in this situation by using a chemical dust or spray that will make the blossoms "set," so your early blossoms will develop into tomatoes.

Excessive heat in summer: high nighttime temperatures of over 57°F. will stop the setting of fruit, and hot dry winds can also cause the blossoms to drop. That's why it is a good idea to buy seedlings, or recommended tomato seeds, from local garden centers. They know the types that best withstand your climate.

Cool, humid conditions: if these prevail, try vibrating the tomato stakes to shake down pollen. Tomato flowers require pollination to set fruit, and cold and damp do not favor pollination. Too much love: too much watering, too much fertilizing in a garden-grown tomato results in lots of vine, but not so much fruit, especially if your fertilizer has too high a percentage of nitrogen. Nitrogen merely makes healthy leaves.

Plants grown in containers require more frequent watering and more fertilizing, since the roots are confined to a small footage of soil. But those growing in the open garden have lots of soil around from which to draw moisture and nutrients, so don't overdo it.

Yellowing leaves are not the death knell; and a single hornworm doesn't mean there's anything wrong with the plant that removing the hornworm won't cure.

Rotate crops. This applies to tomatoes growing in the open garden. It's a good idea to change the locale of all plants seasonally except big ones, such as fruit trees and shrubs, which have to stay where they are.

'Burpee's FV,' a new verticillium and fusarium wilt-resistant hybrid.

Tomatoes growing in garden soil draw certain nutrients from that soil, depleting it as far as tomatoes are concerned, for next year's crop. You can remedy this with fertilizer, but it is also true that the diseases to which plants are subject are often nurtured in the soil in which the plant grows. This means that tomatoes grow best in soil which hasn't grown tomatoes for several years, if the new site has enriched soil.

And finally, if you run into tomato diseases, be sure next year to select only varieties advertised as free of those diseases.

None of this should seem more discouraging than the statement that if you are subject to colds, avoid the conditions that create them. Tomatoes are not difficult to grow, and generally have few problems. The best way to ward off potential problems is to plant healthy seedlings in well-fertilized soil, supplied with lots of moisture-holding humus, and to expect the best, not the worst.

There's time enough to deal with the worst when it appears.

Pests and Diseases

There is a handful of pests and diseases to which tomatoes are susceptible. In some areas of the country they are more susceptible than others.

The most common problem is bloom end rot. The symptom is a dull buff scar or scab on the branch end (as opposed to the stem end) of the fruit. It appears in young tomatoes as well as older ones, and is usually caused by sudden changes of moisture levels in the soil. It doesn't make the fruit inedible, but it doesn't make it appetizing. You are most likely to run into it when plants growing quickly with lots of moisture in the soil are suddenly hit by neglect, or a seasonal drought in your absence. To avoid changes in soil moisture, mulch your plants.

Another cause for blossom end rot is lack of calcium in the soil. To offset this, add a little (5 pounds to 100 square feet) pulverized limestone to the top 12 inches of the garden soil. The high nitrogenous fertilizers, which cause too-rapid growth, also cause this condition.

Fusarium wilt is another tomato problem. The symptoms are yellowing leaves along the lower branches of the plants, which wilt and eventually die. Crop rotation helps ward off fusarium wilt. For the next season, wilt-resistant varieties should be planted. There are lots of them.

Leaf roll is a problem that may nag you, but won't harm you. Symptoms are leaves that for no apparent reason roll up. It won't affect the quality of the fruit or the yield of the plants.

Catface is something you'll recognize at once if you ever run into it. It's an abnormal fruit formation at the blossom end of the fruits and it does look a little like the smirk of a pussy cat. The fruit tastes fine anyway, but doesn't look pretty. To offset it, grow varieties recommended for your locality.

Early blight is more of a problem. The fruit shows brown irregular target-patterned spots. Apply Maneb to the fruit every 5 to 7 days during the growing season.

Late blight is worse yet. Fruits have dark, water-soaked, greenish black blotches. Apply Maneb every 5 to 7 days of the growing season.

Anthracnose is a disease that results in fruit that has water-soaked spots that are round. This more often attacks unstaked plants than staked ones. Apply Maneb weekly, starting when the plants begin to bloom.

Hornworm is a pest, a greenish caterpillar several inches long. It is actually the great lovely Luna moth. If you pick it from the tomato plant that it is chewing up and place it in a closed box with a few air slits, it will soon form a chrysalis; then put it in the light inside the house and it will emerge as a thing of glory. Whether or not you want it around to make more hornworms next year is up to you. Hornworms can do an extraordinary amount of damage to a single tomato plant in a very short time. I pick them off by hand (wearing gloves—they're big and scary) and dispose of them. Commercial growers control them with applications of Carbaryl.

Mites and red spiders are tiny things that create white spots on otherwise healthy leaves. Malathion applied to the underside of the leaves is the most common method of control.

The flea beetle is another sometime tomato pest. Symptoms of its presence are tiny holes in the leaves. Carbaryl or rotenone applied at weekly intervals are the common controls.

Nematodes, or eel worms, in the soil cause swelling or galls on the roots. Once present in the soil, they persist from season to season. Be sure that seedlings do not have the swellings, and avoid soil that has affected tomato plants until the soil has been fumigated for two years.

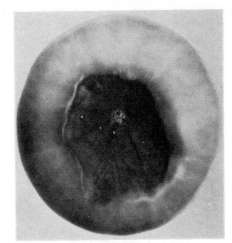

Blosson end rot.

Fusarium wilt.

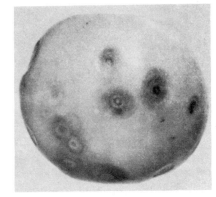

Anthracnose.

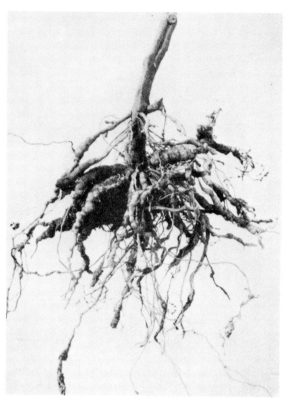

Nematodes.

It goes without saying that rabbits don't belong in the garden, and fencing with chicken wire (the square pattern is handsome) is the only means of keeping them out, if they are neighborly. Slugs may invade and, according to garden legend, pie tins filled with beer will attract and drown them, or a ring of marigolds around the garden will offend them. Slug powder sold at hardwares and nurseries is probably more scientific in dealing with these slimy pests.

The most persistent enemy can be the elusive mole, especially if you have a garden beside uncultivated grounds—a field or woods. They flourish under mulch (though Ruth Stout, that most famous of all gardeners, says she's never been bothered). In no time, moles will eat all your roots, so remove the mulch to dry the ground. Then restore the mulch. And it's also possible to plant garlic cloves every foot along the edge of the garden on the field side, and moles are supposed to stay away.

7
The Cream of the Crop

Harvesting and Storing

Harvest tomatoes reach the stage called "dead ripe." That is, they are full in color, plumb and shiny as satin, with no green or white streaks left around the stem end of the fruit.

Ripe tomatoes of most varieties can stay on the vine for three days before they begin to deteriorate. The right time to harvest them (and all other vegetables, except perhaps some root crops) is just before you are ready to use them. The moment the fruit is picked, it begins to lose some of its nutritive qualities, and this is reflected in the loss of flavor.

Home-ripened tomatoes will still be delicious even if picked well ahead and stored in the refrigerator, but won't be quite so delicious as vine-ripened ones. If you are going away for a holiday, and some tomatoes are almost ready and will obviously ripen long before you return home, pick them and ripen them indoors.

It is easier to tell when the red varieties are dead ripe than the paler varieties. A truly ripe tomato has a strong color, even the yellows and orange-goldens.

None of this pick-and-use advice applies as the tomato-growing season comes to an end. When the first frosts threaten, you can still mature fruit on the vine if you are planning the type of frost protection described on page 46. Plants in containers can, of course, be brought inside to a sunny location. However, if you aren't going to protect the plants through the first frosts, you'll have to pick them.

Tomatoes that are somewhat red will ripen indoors. Store them in good light, but not direct sunlight, unwrapped, between temperatures of 65° to 75°F. Leave space between the fruits; don't set them so that they are touching. A likely place is a north window. Ripened after six or seven days off the vine, they will retain much of their food value and be delicious. Once ripe, store in the refrigerator.

The McDonald garden at its peak, not long before the first frost may strike, with welcome visitors.

Tomatoes that are the color of whitened green—green turning white—will also ripen off the vine. Store as described above. Or wrap loosely in newspaper in a box, if there are lots of them, and store in a bright room until ripe.

Don't wrap any flawed tomato, especially those with brown spots. They won't ripen, only rot. Cut out the brown spots and make piccalilli or other green tomato recipes. (There are dozens of ways of making them!)

Tomatoes that have no white (they are really green) usually will not ripen indoors. Use these as soon as picked. There are recipes for green tomato dishes in the last section of the book.

Beating the Early Frosts

In most areas of the country, a few early frosts threaten tomatoes, and then the weather warms again. If you can get your plants through those cold nights, they will still be ripening fruit when Indian summer comes.

Light frosts that leave hoarfrost on the vines don't actually kill the plants. However, when the sun comes up the next morning, the frost acts as a magnifying glass for sun rays, and leaves and fruit are "burned." Burned fruit usually rots before ripening, and the leaves wilt and dry out. To avoid this type of damage, provide yourself with old sheets, blankets, or large expanses of polyethylene film which can be tossed over the plants on nights when danger threatens.

A more professional approach is to use polyethylene film (which can be saved from year to year) to create permanent nighttime tents for the staked tomatoes. Coverings and film must be removed during the day as the high heat that builds up inside the covering can harm the plants.

When light frosts are over, and Indian summer is going, dig your tomato plants, shake the soil loose from their roots and hang them upside down, fruit still on the branches, in a cool dark place, such as a basement, at 50° or 60°F. A lot of the green fruit will mature.

Part II
Cooking, Canning, and Freezing the Tomato Crop

Tomatoes Vinaigrette.

8
Bumper Crop Recipes

There are lots of nice things to do with tomato crops, be they large or small.

When the first cheery little globes of red appear on bushes bearing the tomato miniatures, use them to brighten green salads, to add a last-minute dash of color to simmering soups and stews, as garnishes around roasts and grilled meats. Poke toothpicks into them and serve as delectable hors d'oeuvres with cheese or sour cream dips. They can replace the sour little onion in a dry martini. Or pickle them in oil-and-vinegar dressing overnight in the refrigerator and serve them as a glamorous appetizer.

As the first flush of ripening tomatoes gives way to a bumper crop, you can prepare lots of tomatoes at one sitting with the recipes that follow.

Since many of the recipes in this section require that the tomatoes be scalded and skinned before using, let me explain the process just in case you haven't had much to do with tomatoes before. Dip the tomatoes in boiling water for ½ minute, cool them under running water for ½ minute, then with a sharp knife, break the skin, and tug at it. If the tomatoes have been scalded long enough (some may take as much as a whole minute), the skin will pull off easily in big patches (like peeling an overdone suntan).

Some of the recipes in this chapter call for canned tomatoes, or for tomato paste: recipes for canning tomatoes, tomato paste, tomato sauces, tomato juices, and other useful preserves, appear in the last two chapters of the book.

Tomatoes with Potato Salad.

Tomato Soup

A hot, tangy soup that makes a meal in itself

10 large, ripe tomatoes	2 or 3 tsp. salt
2 cups cold water	¼ cup carrot, scraped and grated
½ cup fresh parsley, minced	¼ cup green pepper, seeded and shredded
1 medium onion, peeled	¼ cup celery, grated
1 bay leaf	Sour cream to taste
1 tsp. peppercorns	
4 whole cloves	
1 tsp. soy sauce	
2 slices lemon, peeled	

Wash, scald, peel, stem, and quarter the tomatoes. Place them in a large soup kettle with the water, parsley, onion, bay leaf, peppercorns, cloves, soy sauce, lemon, and salt. Over medium heat, stirring occasionally, bring the mixture to a rapid boil, then lower the heat and simmer 20 minutes more.

Strain out the bay leaf, peppercorns, and cloves, then a few cups at a time, run the soup through the blender.

Serve hot with dollops of sour cream.
Yield: 8 portions.

Gazpacho Luncheon

8 ripe tomatoes	1 clove garlic, peeled and minced
2 large cucumbers	Salt and pepper to taste
2 medium onions	
2 green peppers	
2 cups tomato juice	
2 Tbs. wine vinegar	

Wash, scald, skin, and chop the tomatoes. Peel and chop the cucumbers. Peel and shred the onions finely. Stem, seed, and shred the green peppers.

Mix the vegetables together in a large bowl, and stir in the tomato juice, vinegar, garlic, salt, and pepper.

Cover, and chill overnight in the refrigerator.
Yield: 8 generous portions.

Tomato Rarebit

A light dish for summer luncheons

4 Tbs. butter or margarine	⅔ tsp. baking soda
4 Tbs. all-purpose flour	4 cups medium-sharp cheddar cheese, diced
1 cup half-and-half or cream	4 eggs, slightly beaten
1½ cups canned tomatoes, drained; or fresh tomatoes, peeled, cooked soft, drained	½ tsp. mustard
	Salt and pepper to taste
	8 slices sandwich bread, toasted

Melt butter in a chafing dish and stir in the flour very quickly. As soon as it makes a smooth paste, stir in half-and-half or the cream, and beat quickly until the mixture smooths out and thickens.

Make sure the tomatoes are well drained.

In a bowl, mix tomatoes lightly with the soda, then turn them into the cream sauce and combine gently over low heat. Add the cheese, eggs, mustard, and seasonings to taste. As soon as the cheese melts, turn the rarebit onto the bread slices, and serve at once.

Yield: 8–10 generous portions.

Tomatoes with Eggs

Makes a hearty breakfast, served with hot rolls

8 strips of bacon	8 eggs
8 large, ripe toma-toes	1½ tsp. salt
	⅛ tsp. pepper
2 Tbs. olive oil	1 tsp. fresh pars-ley, minced
1 large clove garlic (optional)	

In a large skillet, sauté the bacon until crisp. Remove and drain.

Wash, stem, and core the tomatoes and cut into thirds.

Turn the burner to medium, add the oil to the bacon skillet and set the tomato slices, cut side down, into the skillet. Cook 5 minutes, then turn.

Crush the garlic in a press, or mince it very finely: Press a little garlic onto the cut side of each tomato slice.

Break the eggs into the skillet between the tomato slices. Sprinkle with salt and pepper. Cover the skillet with a lid, or a large plate. Turn heat to low, and simmer until the eggs are set, 5 to 6 minutes.

Use a spatula to lift eggs and tomato slices onto 8 plates. Top each portion with a strip of cooked bacon. Garnish with parsley.

Yield: 8 portions.

Tomato Fry

A quick and easy vegetable side dish

4 large, firm, ripe tomatoes	½ tsp. dried basil or 1 tsp. fresh basil
½ cup all-purpose flour	

Decorated Tomato Juice.

1 tsp. salt	2 Tbs. olive, or
⅛ tsp. pepper	vegetable, oil

Wash, stem, and core the tomatoes, and slice them into rounds ½ inch thick. Discard small end pieces, or save for a salad.

Dredge the slices in the flour, mixed with the salt, pepper, and basil.

In a large, heavy skillet over medium heat, heat the oil for 3 or 4 minutes, then add the tomatoes slices. Sauté slices on both sides until lightly browned. Serve at once.

Yield: 4 portions.

Fresh Tomato Aspic.

Tomato Aspic

Make this aspic in a ring mold with a hollow center, turn it out onto a serving dish, and fill the center with your favorite meat or seafood salad.

3½ cups tomatoes	1 tsp. dried tarra-
1 tsp. salt	gon
½ tsp. paprika	2 Tbs. plain gelatin
1½ tsp. sugar	½ cup cold water
2 Tbs. lemon juice,	½ cup green pep-
strained	pers, chopped,
3 Tbs. onion,	seeded
chopped	½ cup carrots,
1 bay leaf	grated
4 stalks celery with	
leaves	

Over low heat, simmer for 30 minutes the tomatoes, cored, scalded, and skinned, with the salt, paprika, sugar, lemon juice, onion, bay leaf, and the celery, chopped. Strain through a coarse sieve.

Soak the gelatin in the water and stir it into the strained hot tomato juice. Measure the juice and add enough additional water to make 4 cups of liquid. Pour into a mold and chill until it is almost set, then mix in the chopped peppers and carrots.

Chill the aspic until firm. Unmold before serving.

Yield: 6 portions.

Broiled Tomatoes

Great with roasts and broiled meats

4 large, ripe to-	½ tsp. pepper
matoes	Olive, or vegetable,
½ tsp. salt	oil

Heat the broiler, set rack two-thirds of the way to the top.

Wash, dry, stem, and core the tomatoes, and slice in half. Sprinkle cut sides with salt and pepper. Set on broiler rack, cut sides up, and brush tops with oil.

Broil until tomatoes are soft all the way through. If cut sides are becoming too dark, lower heat for last portion of cooking. Broiling should take a total of 10 to 15 minutes.

Yield: 4 or 8 portions.

Stewed Tomatoes

A colorful vegetable dish that is quick and easy

6 large, ripe toma-	2 tsp. light brown
toes	sugar
1 tsp. onion, peeled	1 tsp. parsley,
and minced	chopped
½ cup celery,	1 Tbs. butter or
minced	margarine
⅛ tsp. ground clove	½ cup bread
¾ tsp. salt	crumbs
¼ tsp. paprika	

Wash, scald, skin, stem, and quarter the tomatoes.

Place them in a heavy skillet over low heat and cook, stirring for 20 minutes. After fifteen minutes, add the other ingredients, and stir together gently for the remaining cooking time.

Yield: 6 portions.

Baked Tomatoes

Delicious—the sugar does great things to acidy tomatoes.

6 large, ripe toma-	¼ tsp. pepper
toes	2 Tbs. butter or
3 heaping Tbs.	margarine
light brown sugar	¼ cup canned
1½ tsp. salt	mushroom soup

Preheat the oven to 400°F.

Wash, stem, and core the tomatoes. Stand them on their tops and, with an apple corer, scoop deep, narrow holes downward into the tomato flesh.

Mix the brown sugar, salt, pepper, and butter or margarine into a paste, and push a little of the paste down into the bottom of each gash in the tomatoes. Fill the gashes with mushroom soup.

Place the tomatoes in a greased, shallow baking dish, and set in the preheated oven. Bake for 15 to 20 minutes, or until tomatoes are soft.

Yield: 6 portions.

Ratatouille Mougins

A pleasant variation of a famous vegetable mélange

⅓ cup olive oil	2 medium zucchini
3 large cloves gar-	3 Bermuda onions,
lic, peeled	sliced
6 large, ripe toma-	2 tsp. salt
toes, stemmed	¼ tsp. pepper
1 medium egg-	½ tsp. fresh basil
plant, peeled	¼ tsp. thyme
2 sweet green pep-	
pers, stemmed	
and seeded	

In a very large, heavy kettle, over medium heat, warm the oil. Slice the garlic into the oil.

Wash, stem, and core the tomatoes and cut into 2-inch chunks. Add to the oil and mix well. Cut the eggplant and the peppers into chunks smaller than

Stuffed Tomatoes with Anchovies.

the tomato chunks, and add to the kettle. Stem the zucchini, cut into chunks the size of the tomato chunks, and add. Peel the onions, slice into rounds ¼ inch thick, and add to the mix. As soon as all the vegetables are in, toss all together once. Toss once more after adding salt, pepper, basil, and thyme.

Cover, and simmer at medium heat for 30 minutes. Cook uncovered, stirring occasionally, 30 minutes more, or until the thin, watery sauce has thickened and the mixture has become almost a purée.

Yield: 8 portions.

Tomatoes with Basil

A favorite Provençal way with tomatoes

6 large, ripe toma-	1 tsp. salt
toes	⅛ tsp. pepper
3 Tbs. olive oil	½ cup fresh basil,
1 Tbs. wine vinegar	finely minced

Wash, stem, and core the tomatoes,

Italian Tomato Salad

Use medium-large tomatoes if plum variety isn't available.

18 small plum to- matoes	½ tsp. dried basil, or 1 tsp. fresh, minced
3 Tbs. olive, or vegetable, oil	¼ tsp. pepper
1 Tbs. vinegar	1 tsp. parsley, minced
½ tsp. salt	

Wash, dry, and stem tomatoes. Cut plum tomatoes in half.

In salad bowl, combine the remaining ingredients. Toss the tomatoes in the dressing, and chill for 1 hour before serving.

Yield: 6 portions.

Stuffed Cherry Tomatoes

Use these as an appetizer or an hors d'oeuvre.

16 cherry tomatoes	1 small can crab- meat, lobster- meat, or minced clams
½ tsp. salt	
⅛ tsp. pepper	
3 Tbs. mayonnaise	
¼ tsp. dry mustard	2 tsp. parsley, fine- ly minced
1 tsp. lemon juice, strained	

Wash, dry, and stem tomatoes. Slice off stem ends, and scoop out and reserve pulp.

Drain the pulp, and turn it into a small mixing bowl. Add salt, pepper, mayonnaise, dry mustard and lemon juice; and mix well.

Open and drain the shellfish, and mix thoroughly with the tomato mixture. Stuff cherry tomatoes. Set them on a serving dish, and garnish with minced parsley.

Yield: 4 portions.

Stuffed Tomatoes with Bacon and Other Seasonings.

and slice into rounds ½ inch thick. Place the rounds on a large serving plate.

Combine oil, vinegar, salt, and pepper in a blender. Blend a few seconds and pour over tomatoes; make sure each slice is dressed. Sprinkle minced, fresh basil evenly over each tomato.

Yield: 6 portions.

Stuffed Tomatoes, Italian Style

Serve as a luncheon course, or as a side dish with dinner.

6 medium tomatoes	1½ tsp. capers
½ tsp. salt	1 Tbs. light vege- table oil
¾ cup white rice, cooked	
3 Tbs. fresh mush- rooms, minced	

Preheat oven to 350°F.

Wash, dry, stem, and core the tomatoes. Cut a slice from the top of each, large enough to scoop out the interior. Discard tomato pulp, reserve the tops, and sprinkle salt over interiors of tomatoes. Turn tomatoes upside down to drain.

In a medium bowl combine remaining ingredients, and mix well. Stuff tomatoes with the mixture, and set in a greased baking dish. Place tops on tomatoes, and brush with oil.

Bake 15 to 20 minutes in preheated oven, or until tomatoes are cooked through.

Yield: 6 portions.

Stuffed Tomatoes, American Style

Use these as a main course.

6 medium-large tomatoes	⅛ tsp. pepper
1 lb. ground beef	1 Tbs. curry powder
1 small onion, peeled and minced	1 cup half-and-half or milk
3 Tbs. butter or margarine	½ cup bread crumbs
3 Tbs. all-purpose flour	2 Tbs. butter or margarine
1 tsp. salt	¼ cup parsley, minced

Preheat the oven to 350°F.

Wash and stem the tomatoes, and cut a slice from the top of each, large enough so you can scoop out the interiors. Scoop out and discard pulp, reserving the tops.

In a medium bowl, mix the ground beef with the onion. Over high heat in a medium skillet, melt 1 tablespoon of the butter or margarine. Press the beef into the bottom of a skillet, and brown thoroughly, about 5 minutes on each side. Lower the heat, and break the meat into very small pieces. Mix in flour, salt, pepper, curry powder, and half-and-half or milk. Cook, stirring constantly, until mixture boils and thickens. Spoon into tomato cups.

Grease a shallow baking dish, and set the tomatoes in it. Sprinkle the tops with bread crumbs and dot with 2 tablespoons

Baked Tomatoes with Stuffed Mushrooms.

butter or margarine.

Bake in preheated oven 20 minutes, or until crumbs are golden brown.

Turn the broiler on high, set the tomato tops on the stuffed tomatoes, and broil 2 or 3 minutes, just enough to give the tops a grilled look. Garnish with minced parsley before serving.

Yield: 6 portions.

Steak with Tomatoes

A quick way with a summer meal

4 Tbs. butter or margarine	1 tsp. dried oregano
1½ lbs. chuck steak in 1-inch cubes	1 tsp. Worcestershire sauce
4 large, ripe tomatoes	½ tsp. salt
2 green peppers, seeded and cut into strips	⅛ tsp. pepper
	4 slices Italian bread, 1 inch thick

In a large, heavy skillet, over medium heat, melt the butter or margarine, and brown the meat quickly on all sides, about 3 minutes.

Remove meat to a serving bowl and keep warm. Wash and stem the tomatoes, chop them, and add to the skillet. Add the peppers. Mix in oregano, Worcestershire sauce, salt, and pepper, and cook 5 minutes, or until peppers are softened.

Return the meat cubes to the skillet, mix well with the sauce, and heat through, about 2 minutes.

Set bread slices on dinner plates, and pour meat and sauce over them. Serve hot.

Yield: 4 servings.

Meat Loaf with Tomatoes

Hearty, delicious, easy, and inexpensive

2 slices white bread	¼ tsp. black pepper
2 lbs. hamburger or ground round	4 small, ripe tomatoes
1 cup half-and-half or milk	1 sweet green pepper
2 whole eggs	1 medium onion, peeled
1 small onion, peeled	
2 tsp. salt	

Preheat oven to 400°F.

Place the bread in a large bowl, and pour the half-and-half or milk, over it. Allow to soak 5 minutes, then, with two forks, break the bread into small bits.

Add the meat to the bowl, and mix thoroughly with the bread. Make a well in the center of the meat, and break the eggs into it. Beat the eggs slightly. Grate the onion over the eggs, and add salt and pepper. Beat the eggs into the meat thoroughly.

Grease a loaf pan, or a medium casserole. Shape the meat into a loaf smaller than the baking dish.

Wash, stem, and core the tomatoes, and cut into halves. Stem and seed the pepper and cut into 4 parts. Cut the medium onion into 4 wedges. Press tomato, pepper, and onion pieces in a decorative pattern on the top and into the sides of the meat loaf.

Place the loaf in the preheated oven, and bake for 30 minutes. Vegetables should be just slightly underdone.

Yield: 8 portions, or more.

Tomato Sauce, Italian Style

Serve over spaghetti, rice, noodles—or use in making lasagna.

4 Tbs. olive, or vegetable, oil	1 tsp. salt
2 medium onions, peeled and chopped	⅛ tsp. pepper
	1 tsp. sugar
1 clove garlic, peeled and minced	6 medium-large, ripe tomatoes
	⅔ cup (6-oz. can) tomato paste

In a heavy skillet, over medium heat, warm the oil, and sauté the onions until golden brown. Add garlic, salt, pepper, and sugar. Stem the tomatoes, cut them into small chunks, and toss into the onion mixture.

Sauté until the tomatoes have softened, about 15 minutes, then stir in tomato paste.

Simmer over low heat about 40 minutes, or until the mixture has thickened and the tomatoes have disintegrated.

Yield: About 4 cups of sauce.

9
Canning and
Canning Recipes

Having faithfully followed all the growing instructions and successfully brought in your tomato crop, the next thing that happens is: You have too many.

Can you grow tomatoes too successfully? Yes, of course you can! It happens to most of us sometimes. The season is too perfect and all the plants yield record crops. Or, you didn't trust the estimate that eight plants were enough for you. Or, you became fascinated by the many varieties and planted one of each. Or, all the neighbors and relatives you were going to give tomatoes to—lovely, big baskets of tomatoes—went to the seashore. And there you are, having had tomatoes with eggs for breakfast, and tomatoes with tuna fish for lunch, a Bloody or Virgin Mary for cocktails, and tomato shishkebabs for dinner, and you never want to see a ripe tomato again. (That feeling is short-lived, because vine-ripened tomatoes are *so* good!)

You can put them up lots of ways: Stew them whole and can them; make them into juice and can or freeze it. You can make tomato sauces and can or freeze those. You can make relishes and casseroles and all sorts of good things.

Tomato Canning Recipes

A bushel basket of average-size tomatoes will fill between 15 and 20 quart jars. The weight of a bushel of tomatoes is about 53 pounds. The same quantity of tomatoes, juiced, will fill between 12 and 16 quart jars. This will give you a notion of how many tomatoes you need if you want to put by a lot of tomatoes.

You don't have to save up a bushel of tomatoes to work with. The happiest way to can is to can in small batches. It takes about 2½ to 3½ pounds of tomatoes to fill one quart jar, and about half that to fill a pint jar with tomato juice.

I like to work with enough tomatoes to fill 6 quart jars when I'm making whole, canned tomatoes, because that's the number of jars my canner holds at one time. To fill 6 quart jars requires between 15 and 20 pounds of tomatoes. To fill 6 quart jars with tomato *juice* usually takes 18 to 24 pounds of tomatoes. I find that to can 20 pounds of tomatoes I spend about 2 hours, as much time as I want to spend at any one time on canning.

Canning isn't difficult, but it is exacting. You cannot can in slipshod fashion. If you follow canning instructions exactly, you need never fear botulism or other forms of poisoning or spoilage. Spoilage occurs when canned foods are improperly processed, handled, and stored. Botulism is not a threat in the processing of tomatoes, a high-acid food.

Canning Equipment

When we talk about "canning" we are discussing one of three types of canning procedure.

The method used for most vegetables is to can using a pressure canner or cooker for the processing period. Processing means heating for a period of time to kill harmful bacteria.

The method used for canning most fruits, and this includes tomatoes, is to process the foods in a boiling water bath. The third method for canning foods, is to precook the food, then to pack it into sterile jars, and to seal the jars with paraffin. Paraffin-sealed jars are not processed. Some foods are precooked, and *also* processed. These are sealed with regular canning lids and caps before processing. The method is called "open-kettle canning" because you precook the foods usually in a kettle without a lid, stirring as it cooks.

If you get involved with canning, you'll hear a lot of talk about "low-acid" vegetables and "high-acid foods." Foods low in acid content must be processed at temperatures of 240°F. which, in a pressure cooker or canner, reads as 10 pounds pressure at sea level to 2,000 feet altitude. At this high heat under pressure, the potentially harmful bacteria in low-acid foods are destroyed. It takes

a pressure cooker or canner to maintain the right temperatures.

A boiling water-bath canner which is really just a great big kettle, can only maintain temperatures of 212°F. The boiling water-bath canner method is suited to all the high-acid foods, which include fruits and tomatoes. Acids (vinegar and lemon juice, for instance) discourage harmful bacteria, and so foods high in acid content can be processed at less lethal temperatures.

About altitudes: Because water boils at lower temperatures, the lower the atmospheric pressure (that is, at higher altitudes), processing times for canning recipes must be increased when canning at altitudes above 2,000 feet. The following tables show by how many minutes, in the case of boiling water-bath canning, and by how many pounds pressure, in the case of pressure canning.

To sum it up, you need a pressure canner or cooker to can low-acid foods, which means most vegetables. And you need a boiling water-bath canner to can high-acid foods, which include tomatoes, as well as most fruits.

Commercial manufacturers offer several types of boiling water-bath canner. Basically, they are all big, very deep kettles equipped with tight-fitting lids and racks on which canning jars can sit during the processing period. The lid is essential to keep the water boiling evenly during the processing period. The rack is essential to keep the glass jars from resting on the floor of the kettle where temperatures may become high enough to harm them. You may have a big kettle suitable for use as a water-bath canner. To test its suitability, set a rack an inch high in the bottom and rest quart, pint, or half-pint jars on the rack. If there is enough room above the jar tops for 1 to 2 inches of water, plus enough room for the water to boil without spilling, and if you have a tight-fitting lid for the kettle, it can be used. Canning in half-pint size jars isn't very satisfactory unless you really want cupful batches.

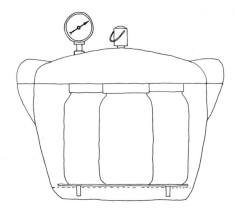

Pressure canner, shown in cross-section, can be used for processing tomatoes under pressure or in boiling water bath. For boiling water-bath canning, leave the petcock open so the steam can escape freely.

Left: When sealing this type container, put the flat metal lid on the rim of the jar with the sealing side on the glass. Hold the lid down and screw the band on firmly. Remove the screw band after the seal has been tested, by tapping, as described on page 62. Right: This type equipment is called "self-sealing." Moisten the rubber ring and place it on the glass lid. Place the lid on the jar with the rubber against the rim of the jar; then put the screw band on and screw it down tightly. Then turn the screw band back ¼ inch to loosen it a little. After processing, and as soon as you remove the jar from the canner, screw the band down tightly to complete the seal.

Altitude Chart for Pressure Canning

Altitude	Process at
Sea level to 2,000	10 pounds
2,000 to 3,000 feet	12 pounds
3,000 to 4,000 feet	12 pounds
4,000 to 5,000 feet	13 pounds
5,000 to 6,000 feet	13 pounds
6,000 to 7,000 feet	14 pounds
7,000 to 8,000 feet	14 pounds
8,000 to 9,000 feet	15 pounds
9,000 to 10,000 feet	15 pounds

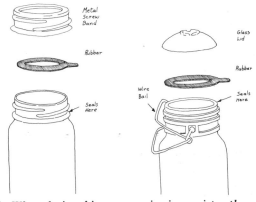

Altitude Chart for Boiling Water-Bath Canning

Increase Processing if the Time Called for Is:

Altitude	Up to 20 min.	More than 20 min.
1,000 ft.	1 min.	2 min.
2,000 ft.	2 min.	4 min.
3,000 ft.	3 min.	6 min.
4,000 ft.	4 min.	8 min.
5,000 ft.	5 min.	10 min.
6,000 ft.	6 min.	12 min.
7,000 ft.	7 min.	14 min.
8,000 ft.	8 min.	16 min.
9,000 ft.	9 min.	18 min.
10,000 ft.	10 min.	20 min.

Left: When closing this type canning jar, moisten the rubber ring and place on the shoulder of the jar. Screw the cap down very tightly over the wet ring. Then turn the screw cap back about ¼ inch to loosen it slightly. Process the foods as directed. When you remove the jar from the canner, immediately screw the cap down tightly to complete the seal. Right: The lid on this type canning jar is held down by the wire bail. Moisten the rubber ring and put it on the screw thread at the top of the jar, and press the glass lid on the rubber ring. Then put the lower bail wire over the top of the glass lid so that it fits into the groove on the lid. Push the other bail wire down against the side of the jar. The jar is now sealed. Process the jar, and do not disturb the bail wire after processing.

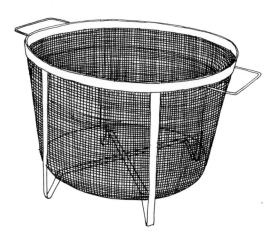

Wire basket makes scalding tomatoes easy and quick, but isn't essential equipment. A metal colander will do as well, but takes longer to heat up.

Not essential, but handy, in canning, are these four pieces of equipment, found in most hardware stores: a cup-shaped funnel to make pouring tidy; a ladle with a lip; a footed colander for draining washed and scalded fruit; a food mill to make it easy to mash pulp. For canning tomatoes, which are acid in content, it is best to work with glass or enamel-covered equipment. In contact with iron, tomatoes may darken.

A pressure cooker (illustrated on page 72) can be used as a water-bath canner, if large enough to hold jars of the size you wish to can in, as long as it meets the depth requirements described in the paragraph above, and as long as you leave the petcock open so the steam can escape, and no excessive pressures build up.

The rest of the equipment needed for canning tomatoes is simple enough, and you probably have most of it on hand: Jars may be the exception. For canning tomatoes whole, I use wide-mouth Mason jars with wide-mouth Mason screw caps and dome lids, and I put up in quart sizes. For small batches of sauces, I use wide-mouth pint jars with similar caps and lids. For pickles and relishes meant for use as gifts, I often use half-pint jars.

I also use some inherited quart jars of the type that has a rubber ring held in place with a dome lid secured by a wire bail. The illustrations on page 59 show the various types of jars offered commercially, and explain how each is sealed.

The jars used for canning are very important. They must be strong enough to withstand the high heats of the water-bath canner. The rims of the jars must be flawless to ensure perfect seal. Before you use either a new or an old jar for canning, run your finger around the top of the rim to make sure no nicks or cracks are there. For recipes such as pickles and relishes to be sealed with paraffin, without heat processing, you can use retreads—peanut butter jars, jam or jelly containers. However, these materials aren't recommended for processing in either a boiling water-bath canner or pressure canner or cooker.

Some of the other items useful during canning include a wire basket (the kind used to drain salad greens, for instance), a sharp paring knife, a long-handled wooden spoon for stirring when open-kettle canning, a stainless steel or porcelain-coated ladle with a pouring lip, a cup-shaped glass or porcelain-covered funnel, a jar lifter, a rack for cooling filled, processed jars. Use glass, porcelain, or stainless steel equipment because acid food interacts with iron and its alloys.

Hot- and Cold-Pack

Before we go into recipes meant to use up all those successful tomatoes, there is one or more bits of information you should have about canning methods. Most manuals for canning discuss the virtues of "hot-pack" and "cold-pack" canning.

Cold-pack means the food to be processed is packed into the jars, raw and uncooked.

Hot-pack means the food to be processed is packed into the jars after partial cooking.

The basic approach to canning is the same in either case. You measure the amount of tomatoes needed to fill the jars your processing equipment handles at one session, prepare these, pack them raw, or else pack them after partial cooking, into their jars, and process them. Obviously, cold-pack foods require longer processing times than hot-pack foods. Cold- or raw-pack tomatoes in quarts are processed for 45 minutes in a boiling water-bath canner: Hot-pack tomatoes are processed, in quart sizes, for 15 minutes. The cold-pack method may be a little easier on the cook—the hot-pack method takes less processing time.

Whole Tomatoes: Cold-Pack

1. Gather enough tomatoes to generously fill the number of jars you plan to process, and a few extras to make juice with which to fill the jars. Select firm, ripe tomatoes, without blemishes and without green streaks by the stems.
2. Assemble all the equipment you will use: the canner, jars, caps, lids, wire basket, paring knife, funnel, measuring cup, spatula, jar lifter, and rack for cooling. A couple of towels can be used if you have no rack.
3. Wash and rinse the jars, caps, and lids in hot, soapy water. Leave the jars in hot water until ready to use them. Pour boiling water over the lids, and leave them in it until you are ready to use them.
4. Wash and drain the tomatoes in cold water.
5. Bring 2 or 3 quarts of water to a rapid boil in a large kettle. Fill the sink with cold water.

One type of jar lifter has scissor-like handles.

This type of jar lifter lifts from top and is a little easier to handle.

6. Half a dozen at a time, set the tomatoes into the wire basket, and scald them in the boiling water for ½ minute. Plunge the basket into the cold water in the sink for about 1 minute. Then with the paring knife, cut out the stem and white core near the stem, and pull away the tomato skin. After scalding, the skin comes away easily in large patches.

7. Pack the tomatoes into the hot jars, and press the tomatoes firmly into place. Each jar should have about ½ inch head room between the tops of the tomatoes and the top of the jar rim.

8. Put the extra tomatoes into a blender, or a juicer, in chunks, and use the juice to fill the tomato jars to within ½ inch of the rim tops.

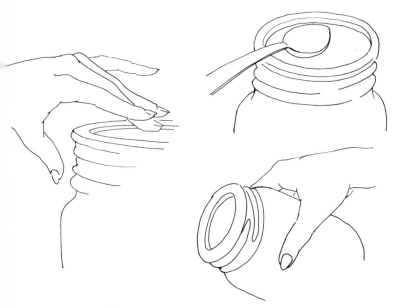

Three ways to test jars to make sure they are vacuum sealed are shown here. Left: With your finger, press the center of the lid after processing and cooling; if it is down and will not move, the jar is sealed. With a spoon: tap the jar with a metal spoon, and it will give a clear, ringing sound if the seal is correct. Bottom: Tilt the jar if you have doubts about the seal: if any liquid trickles through, the seal is not complete. Imperfect seals require that you reprocess the contents, or else use them up right away. An alternate is to freeze the food in the imperfectly sealed jars.

9. To each quart jar, add 1 teaspoon of salt. Use the type of salt advertised as pure, or unrefined. Some other types of salt contain starch which may then turn up in canned goods as a whitish sediment suspended in the liquids, or settled on the jar bottom. It doesn't look pretty, and it always worries you, though starch as sediment isn't harmful to canned goods. If you are canning in pint jars, use ½ teaspoon of salt to each pint jar.

10. If you are canning low-acid tomatoes, such as the little yellow plums, and some of the gentle pinks and oranges, add 1 teaspoon of strained lemon juice to each quart jar, and ½ teaspoon to each pint jar to increase acid content.

11. Fill the water-bath canner half full, and place the rack on the bottom. Over high heat, bring it to a boil.

12. While the water is heating, run a spatula around the interior sides of the filled jars to remove all air bubbles, then wipe the tops and threads of the jar, place the lids and caps on and close them. As you finish each jar, place it in the water.

13. When all the jars are in the canner, fill the canner so that the water level is 1 to 2 inches above the jar-lid tops, put the cover on the canner, and bring the water to a boil. Start counting processing time as soon as the water hits a steady but gentle boil.

14. At altitudes less than 1,000 feet above sea level, process pints 35 minutes and quarts 45 minutes.

15. When processing time is up, remove the jars from the canner with the jar lifter and set them away from cold air or drafts on a cooling rack, or on a couple of folded towels. Allow the jars to cool for about 12 hours.

16. Test the jars for seal. The seal must be perfect, or the tomatoes may spoil during storage. To test, press the center of the lid. If the lid is down and won't move, or if it stays down when

pressed, the jar is probably sealed. If you have doubts, tap the center of the lid with a metal spoon. A clear ringing sound generally means the seal is good. You can also test for vacuum seal by tilting the jar. If there is any leakage, obviously the seal isn't good. After satisfying yourself that each and every jar is properly sealed, remove the screw bands, wash and store them. Wipe the jars one more time, and store in a dark, reasonably cool place. A place below 70° and above 32°F. is suitable. Ideal is probably somewhere between 50° and 65°F.

Whole Tomatoes: Hot-Pack

Handling of tomatoes for hot- and cold-pack differs little. In the above recipe for cold-pack canning of tomatoes, insert a step 6a: After the tomatoes have been skinned and stemmed, place them in a large kettle with the tomato juices rendered, as described in step 8: Bring the tomato juice to a rapid boil, and boil the tomatoes, stirring constantly for 5 minutes. Pour immediately into the hot jars, and proceed as in the recipe for Whole Tomatoes: Cold-Pack.

Process (hot-pack) whole tomatoes in pint jars for 10 minutes. Process quarts for 15 minutes, at sea level to 1,000 feet above.

Tomato Juices

Making tomato juices for canning is pretty easy, and it's an excellent way to use up an overgenerous crop. You can make juice plain or seasoned: In either case, once the juice has been poured into jars, it must be processed, as described in the recipes below, by the boiling water-bath process. Put tomato juice up in pints: One pint is two cups, or four to six servings in small tomato juice glasses.

The steps describing canning procedures in the recipe for Tomatoes Whole on the preceding pages are followed in canning tomato juices. Remember when planning processing times to check the Altitude Chart for Boiling Water-Bath Canning, on page 59.

Tomato Juice

9–12 lbs. very ripe tomatoes	1½ tsp. salt
1 Tbs. lemon juice, strained	

Wash the tomatoes, stem, and core them; place them in a large kettle; and over medium heat, bring them to simmering. Stirring often, cook for about 15 minutes, or until the tomatoes are soft.

Press the tomatoes through a food mill or a sieve to extract juice. Return the juice to the kettle, and bring to a boil. Stir in the lemon juice and the salt, simmer 1 minute more.

Pour into hot jars, leaving ½ inch head room between juice and rim top. Process in a boiling water bath for 10 minutes, at sea level to 1,000 feet.

Yield: About 6 pints.

Savory Tomato Juice

9–12 lbs. very ripe tomatoes (about 18 cups, chopped)	½ tsp. thyme
1 cup celery, chopped	2 tsp. salt
½ cup onion, chopped and peeled	1 Tbs. sugar
¼ cup lemon juice, strained	2 tsp. Worcestershire sauce
	¾ tsp. Tabasco sauce
	1 tsp. soy sauce

Wash tomatoes, stem, and core them. Cut coarsely. Measure about 18 cups, and turn the tomatoes into a large kettle. Add celery and onion, cover and cook over slow heat until tomatoes are soft, about 15 minutes. Stir often.

Press tomatoes through a food mill, or a sieve, to extract juice. Measure about 12 cups of juice and return to the kettle. Simmer gently, uncovered, for about 30 minutes. Add remaining ingredients, and simmer 10 minutes more. Taste, and add more salt if you like. Return to a simmer, then pour into hot jars, leaving ½ inch head space.

Process in a boiling water-bath canner for 10 minutes, at sea level to 1,000 feet.
Yield: About 5 pints.

Tomato Paste and Sauces

Making tomato paste is tedious, but rewarding, because the end product is more flavorful than the tomato pastes offered commercially. If you particularly favor some herbs such as oregano or rosemary in dishes made with tomato paste, you can add these to the recipe that follows. Add about ¼ teaspoon of the herb for each pint of paste. Tomato purée is made much as tomato paste is made, but isn't boiled down to quite such a thick consistency. Tomato sauces are made as purée is made, but contain many more condiments.

Sauces, pastes, and purées are partially cooked, turned into canning jars, then processed in the boiling water bath for times noted on page 59.

Remember that processing time varies with altitude, as shown in the Altitude Chart for Boiling Water-Bath Canning, page 59.

Basic procedures governing preparation of jars, capping instructions, tests for correct seal after processing, must be followed in the next recipes. These basic procedures appear under the recipes for Tomatoes Whole: Cold-Pack, and Tomatoes Whole: Hot-Pack.

Tomato Paste

20–25 lbs. ripe tomatoes (32 cups, chopped)	2 bay leaves
	1 Tbs. salt
	1 clove garlic (optional)
1½ cups sweet red peppers, stemmed and seeded (use green if red aren't available)	

Wash, stem, and chop the tomatoes. Combine with peppers, bay leaves, and salt in a large kettle. Bring to the simmer point, and simmer, stirring occasionally, 1 hour. Press through a sieve.

Return to the kettle, add the garlic, and continue cooking until thick enough to mound on a spoon, about 2½ hours. Stir frequently. Remove the garlic.

Pour into hot jars, and cap. Process in a boiling water bath 45 minutes, at sea level to 1,000 feet.
Yield: 9 half-pints.

(Some varieties offered are especially suited for making tomato paste: among them, 'San Marzano' by Burpee. If you are planning to make a lot of tomato paste and tomato casseroles and soups, plant 3 or 4 of these varieties especially for the purée. They make excellent tomato paste.)

Tomato Purée

10–13 lbs. ripe tomatoes (16 cups chopped)	2 cups celery, chopped
3 cups onions, peeled and chopped	1½ cups sweet green pepper, stemmed, seeded, and chopped
2 cups carrots, scraped and sliced	1 Tbs. salt

Wash, scald, peel, stem, and core the tomatoes; and chop them. Combine all ingredients in a medium kettle, and simmer until tender, about 10 to 15 minutes. Press through a sieve, or blend ½ minute 1 to 2 cupfuls at a time. Return to the kettle, and simmer until thick, about 1½ hours.

Pour into hot jars, leaving ¼ inch head space. Process in a boiling water bath for 45 minutes, at sea level to 1,000 feet.
Yield: 8–10 half-pints.

Tomato Catsup

Catsup is best made with tomatoes that are more meat than liquid. Varieties such as 'Roma' and those called 'Heinz' and 'Campbell' are preferred for catsups, thick sauces, and pastes.

20 lbs. ripe tomatoes (4 quarts chopped)	1 tsp. whole allspice
1 cup onions, chopped	1 stick cinnamon
½ cup sweet red peppers, chopped	1 cup granulated sugar
1½ tsp. celery seed	1 Tbs. salt
1 tsp. mustard seed	1½ cups vinegar
	1 Tbs. paprika

Wash and core the tomatoes, chop and measure 32 cups (4 quarts). In a large kettle, cook the tomatoes with the onion and the peppers until soft, about 20 minutes. Press through a food mill or a sieve, and discard skins. Return to the kettle, and simmer rapidly over medium heat until the volume is reduced by about half. Stir often. This will take about 1 hour.

Tie the whole spices into a cheesecloth bag, and add to the tomatoes with the sugar and the salt. Lower the heat and simmer another 25 minutes, stirring often. Add the vinegar and the paprika, and simmer another 10 minutes, stirring.

Pour into hot jars, cap, and process in boiling water for 10 minutes, at sea level to 1,000 feet.

Yield: About 3 pints.

Tomato Sauce for Spaghetti

8 lbs. tomatoes	2 Tbs. white vinegar
1 tsp. celery seed	
1 tsp. dry mustard	2 tsp. salt
¼ tsp. thyme	1 bay leaf
½ tsp. oregano	
2 cups fresh mushrooms, chopped	

Wash the tomatoes; stem, core, and quarter them. In a large kettle, bring the tomatoes to simmering. Add the celery seed, mustard, thyme, and oregano, and simmer 10 minutes, stirring often. Put through a sieve, or a food mill, or the blender (½ minute), a few cups at a time.

Return to the kettle. Add the chopped mushrooms, vinegar, salt, and bay leaf.

Bring to a boil, and simmer 45 minutes, uncovered. Stir occasionally. The sauce is ready when it has thickened to the consistency of light catsup. Remove the bay leaf.

Pour into hot jars, cap, and process in a boiling water bath 30 minutes at sea level to 1,000 feet.

Yield: About 3 pints.

Barbecue Sauce

(Before using this sauce, add 1 cup of salad oil to each pint jar, and mix thoroughly.)

10–13 lbs. ripe tomatoes (16 cups, chopped)	2 cups onions, peeled and chopped
2 hot red peppers	1 cup brown sugar, firmly packed
2 medium onions, peeled and quartered	1 cup cider vinegar
2 cups celery, chopped	2 cloves garlic, peeled and crushed
1½ cups sweet green peppers, seeded	1 Tbs. unrefined salt
⅛ tsp. cayenne pepper	1 Tbs. mustard
	1 Tbs. paprika
1 tsp. peppercorns	1 tsp. Tabasco sauce

Wash, scald, skin, stem, and core the tomatoes, and chop coarsely. Stem and seed the hot peppers, using rubber gloves. Keep your hands from your face. Combine tomatoes, hot peppers, onions, and sweet peppers. In a medium kettle simmer until the vegetables are soft, about 30 minutes. Press through sieve.

Return the mix to the kettle, and simmer until reduced by one half, about 45 minutes.

Tie the peppercorns into a cheesecloth bag, and add, with remaining ingredients, to the sauce. Simmer until the mixture thickens, about 1½ hours. Stir frequently.

Pour into hot jars, cap, and process 20 minutes in a boiling water bath, at sea level to 1,000 feet.

Yield: 4 to 5 pints.

Tomato Catsup

12½ lbs. ripe to-
matoes
2 medium onions,
peeled
¼ tsp. cayenne
pepper
2 cups cider vine-
gar
1½ Tbs. broken
stick cinnamon

1 Tbs. whole cloves
3 cloves garlic,
peeled and
minced
1 Tbs. paprika
1 cup granulated
sugar
2½ tsp. unrefined
salt

Wash, stem, core, and slice the tomatoes, and in a medium-size kettle over
medium heat, simmer until soft.

Slice the onions into a small kettle.
Add enough water to cover, and boil
until tender, 5 to 10 minutes. Press the
cooked onions and tomatoes through a
sieve, and mix well together. Stir in the
cayenne pepper, return the ingredients
to the larger kettle, and boil rapidly until
the mélange has reduced to half its original volume, about 30 minutes.

Place the vinegar in an enamel or
glass pan. Add the cinnamon, cloves,
and the garlic cloves to the vinegar, and
simmer 30 minutes. Bring to a rapid boil,
cover, and remove from the heat. Allow
this to stand at room temperature, covered, until ready to mix with tomatoes.

When the tomato mixture has cooked
down, stir in the vinegar mixture. There
should be about 1¼ cups. Add the paprika,
sugar, and salt, and boil rapidly until
thick, about 10 minutes.

Pour into hot jars, leaving ½ inch head
room. Cap, and process in a boiling water
bath 5 minutes, at sea level to 1,000 feet.
Yield: 6 pints.

Chili Sauce

10–13 lbs. ripe
tomatoes (16 cups
chopped)
1 medium onion,
peeled (⅔ cup,
chopped)
1½ cups granulated
sugar
¾ tsp. Tabasco sauce

½ tsp. curry pow-
der
2 cups vinegar
5 tsp. unrefined
salt
1 tsp. ground cin-
namon
1 tsp. dry mustard

Wash, scald, skin, stem, and core the
tomatoes, and chop coarsely with onion.
Place in a large kettle, and add all the
remaining ingredients. Over medium
heat, bring the mélange to a boil, and
simmer two hours, stirring frequently.

Pour into hot jars, cap, and process in
a boiling water bath for 5 minutes, at sea
level to 1,000 feet.
Yield: 6 pints.

Marmalades and Sweet Things

Have you ever tasted tomato marmalade? Little half-pint jars make con-
versation-piece Christmas and birthday
gifts. Because tomatoes are acidy, they
combine especially well with sweeteners
to create a sweet-sour spread that de-
lights most everyone.

The sweetened tomato preserves, con-
serves, and marmalades below are processed in a boiling water bath following
the steps described for Whole Tomatoes:
Hot-Pack. But processing times are much
shorter, since the product being processed
has already been cooked.

Remember to check processing times
with the Altitude Chart for Boiling Water-
Bath Canning, page 59.

Tomato Marmalade

7–10 lbs. ripe to-
matoes (12 cups,
chopped)
2 medium oranges
2 medium lemons

10 cups granulated
sugar
6 Tbs. whole cloves
6 Tbs. broken cin-
namon sticks

Wash, scald, and skin the tomatoes,
as described for Whole Tomatoes: Cold-
Pack. Remove tomato stems and cores,
and cut into small pieces. You need 12
cups. Place in a medium kettle.

Stem, slice the oranges and the lemons
into quarters, then slice the quarters into
paper-thin slices.

Pour off the tomato juices, and stir
the sugar into the tomatoes, until dissolved. Mix in the sliced oranges and
lemons.

Tie the spices into a cheesecloth square, and add to the kettle. Place the mixture over high heat, stirring often, and bring to a rapid boil. Turn the heat down, and simmer until the mixture is clear and thick, about 50 minutes. Stir occasionally.

Pour into hot jars, leaving ½ inch head room at the top. Cap, and process in a boiling water bath for 10 minutes, at sea level to 1,000 feet.

Yield: 8 half-pints.

Green Tomato Preserves

11 cups green to- matoes, stemmed, cored, and chopped	8 cups granulated sugar 2 lemons, sliced paper-thin

In a medium kettle, place the chopped tomatoes, cover with boiling water, and simmer 5 minutes. Drain.

Add the sugar to the tomatoes, and let stand at room temperature 4 hours.

Drain the syrup into a small kettle, and over high heat, stirring constantly, boil the syrup rapidly until it spins a thread when dropped from a spoon, about 15 minutes.

Return the tomatoes to the syrup, add the lemon slices, and boil until the syrup is thick and clear, about 10 minutes.

Pour into hot jars, leaving ½ inch head room. Process in a boiling water bath 10 minutes at sea level to 1,000 feet.

Yield: 4 pints.

Tomato Conserve

18 cups ripe toma- toes, stemmed, cored, and chopped 3 tsp. ground ginger	6 cups granulated sugar 3 lemons, sliced paper-thin

In a medium kettle, place the tomatoes and over low heat, stirring occasionally, simmer 45 minutes.

Add the ginger, sugar, and lemon. Simmer until the tomato juice is thick and smooth.

Pour into hot jars, leaving ½ inch head room between the tomato conserve and the top of the jar rims.

Process in a boiling water bath for 10 minutes, at sea level to 1,000 feet.

Yield: 4 pints.

Pickles, Relishes, and Chutneys

Pickled things traditionally have a warm place in the hearts of Americans because many of the immigrants who landed here brought their provisions in kegs pickled in brine or vinegar solutions. Nesting in their new land, they continued to salt down for the winter whatever could be spared from their first crops. Vinegar from apples was the most commonly used pickling solution in the old days.

The procedures for processing pickles are similar to those described in Tomatoes Whole: Raw-Packed or Hot-Packed. The processing method used for pickles is a boiling water bath. You need the equipment and supplies described on those pages, but will probably prefer to put up pickles and relishes in smaller jars, the pint or half-pint sizes called for in the recipes below. Jars must be washed, capped after filling, and tested for seal when processing pickles just as when processing Tomatoes Whole. Remember when planning your recipe, to check the Altitude Chart for Boiling Water-Bath Canning on page 59 if you live above 1,000 feet above sea level.

Pickle recipes generally call for open-kettle cooking followed by water-bath processing. Precooking pickles (hot-pack) eliminates shrinkage. If your pickled tomatoes come out shriveled, chances are you used too strong a salt, sugar, or vinegar solution, or processed too long. Watch timing carefully. Remember to start counting as soon as the water in the canner comes to a boil, and to remove the jars as soon as the time is up.

Green Tomato Pickles: Sweet

10–13 lbs. green tomatoes (16 cups whole)	4 cups granulated sugar
¼ cup unrefined salt	1 tsp. whole cloves
½ tsp. powdered alum	½ tsp. ground cinnamon
2 quarts boiling water	1 Tbs. celery seed
3 cups vinegar	1 Tbs. mustard seed
1 cup water	½ tsp. ground allspice

Wash, stem, and core the tomatoes. Cut them into thin chunks, sprinkle them with the salt, and allow to stand at room temperature overnight.

In the morning, drain the tomatoes and place them in a medium kettle. Measure the powdered alum into the boiling water, and pour the water over the tomatoes. Allow to stand for 20 minutes. Drain. Under the tap, rinse in cold water, and drain thoroughly. Tie spices into a cheesecloth bag, and combine with the vinegar, water, and sugar in a small kettle, and bring to a rapid boil. Boil 1 minute, then pour over the tomatoes. Allow the tomatoes to stand in this solution overnight.

In the morning remove spice bag and bring tomatoes in the solution to a rapid boil, and boil 1 minute. Pack into hot jars, leaving ½ inch head room at the top. Cap, and process in boiling water bath 5 minutes, at sea level to 1,000 feet.

Yield: 8 pints.

Kosher Green Tomato Pickles

15–20 lbs. small green tomatoes (24 cups)	6 cloves garlic
6 whole stalks celery	2 quarts water
	1 quart vinegar
1½ sweet green peppers, quartered	1 cup unrefined salt
	6 large sprigs fresh dill, or 2 tablespoons dill seed

Wash, dry, stem, and core the tomatoes, and pack into 6 hot quart jars. Press the tomatoes down, so each jar is well filled. Leave ½ inch head room at the top.

Into each quart jar, press 1 whole stalk of washed, dried celery, ¼ of a washed, dried, seeded green papper, 1 garlic clove, peeled.

Place the water, the vinegar, and the salt in a medium kettle. Bring to a boil. Add the dill, and simmer 5 minutes.

Pour the hot brine over the pickles in each jar, leaving ½ inch head room at the top. Add 1 dill sprig, if you are using fresh dill, to the top of each jar.

Cap, and process in a boiling water bath for 15 minutes, at sea level to 1,000 feet.

Store the processed jars for 4 to 6 weeks before using.

Yield: 6 quarts.

Chow-Chow Relish

2½ to 3½ lbs. green tomatoes	¼ cup unrefined salt
4 cups cabbage, shredded	1½ cups vinegar
2 sweet red peppers, seeded and shredded	1½ cups water
	2 cups firmly packed light brown sugar
4 cups green peppers, seeded and shredded	1 tsp. dry mustard
	1 tsp. ground turmeric
2 large onions, peeled and minced	1 tsp. whole celery seed

Wash, dry, stem, core, and chop the tomatoes. Mix with cabbage, red and green peppers, the onions sprinkled with salt, and let stand overnight at room temperature.

In the morning, pour off the liquid, and squeeze the vegetables to remove any remaining liquid.

Place the vinegar, water, sugar, and spices, knotted into a cheesecloth bag, in a large kettle, and bring to a boil. Simmer 5 minutes, remove cheesecloth bag, then stir the vegetables into the boiling syrup. Bring back to a full rolling boil, turn into hot jars, leaving ½ inch head room at the top.

Cap, and process in a boiling water bath 5 minutes, at sea level to 1,000 feet.

Yield: 6 pints.

Piccalilli: Green Tomato Relish

12½ lbs. green to-
matoes
8 large onions,
peeled
10 sweet green
peppers, seeded
3 Tbs. unrefined
salt
6 small hot pep-
pers
1 quart vinegar

1 Tbs. ground cin-
namon
1 Tbs. ground all-
spice
½ tsp. ground
cloves
3 Tbs. dry mustard
6 whole bay leaves
1¾ cups granu-
lated sugar

Scald, skin, stem, and core the tomatoes. Chop with onions and green peppers. Sprinkle with salt. Allow to stand
overnight at room temperature.

In the morning, stem, seed, and chop
the hot peppers, wearing rubber gloves;
keep your hands from your face. Drain the
tomato mixture. Mix the hot peppers into
the tomato mixture, along with the vinegar, the spices, tied into a cheesecloth
bag, bay leaves, and sugar. Place over
medium heat, and simmer slowly until
all the vegetables are tender, about 15
minutes. Remove cheesecloth bag.

Pack into hot pint jars allowing 1 bay
leaf to each jar. Cap, and process in a
boiling water bath for 5 minutes, at sea
level to 1,000 feet.

Yield: 6 pints.

Western Relish

½ cup granulated
sugar
1 Tbs. salt
1 Tbs. turmeric
2 16-oz. cans whole
kernel corn
2 cups chopped
ripe tomatoes
2 cups onion,
peeled and
chopped
2 cups cucumber,
peeled and
chopped

2 cups sweet green
pepper, stem-
med and seeded
1 cup celery,
chopped
2 dried red hot
peppers, seeded
and crushed
(optional)
1 cup cider vinegar
2 tsp. mustard
seed
¼ cup cold water
2 Tbs. cornstarch

Combine sugar, salt, and turmeric, in
a large kettle, and add the vegetables,
the vinegar, and mustard seed. Bring to
the boiling point, reduce the heat, and

simmer 30 to 40 minutes, stirring occasionally. Mix water and cornstarch and
add to the hot mélange. Stir and simmer,
until slightly thickened, about 5 minutes.

Pour into hot jars, and process in boiling water bath 15 minutes at sea level
to 1,000 feet.

Yield: 4 or 5 pints.

Tomato Chutney

30 large, ripe to-
matoes
6 large apples
6 large pears
6 large peaches
4 large onions
2 cups cider vinegar

4 cups granulated
sugar
2 Tbs. unrefined
salt
1 package mixed
pickling spices

Wash, scald, skin, stem, and core the
tomatoes, apples and pears; peel and
seed the peaches; peel the onions. Chop
all these together coarsely.

Place all the remaining ingredients in
a large kettle, with the pickling spices
tied into a cheesecloth bag and set over
high heat. Bring to a rapid boil. Add the
vegetables and fruit mélange, bring back
to a boil, and boil until thick, about 20
minutes. Remove the pickling spices.

Pour into hot jars, leaving ½ inch head
room and process in a boiling water bath
5 minutes at sea level to 1,000 feet.

Yield: 12 pints.

Canning Tomatoes Mixed with Other Ingredients

The few vegetables and spices mixed
with tomatoes in the recipes above can
all be processed by a boiling water bath
after being cooked for the time specified
in each recipe. If you want to can tomato
soups and tomatoes mixed with other ingredients, okra or ground beef for instance, the processing method recommended is the pressure canner.

The pressure cooker is mentioned on
page 58 as a warm water-bath canner.
The procedure for canning with a pressure
cooker is very similar to that for canning

in a boiling water bath. The difference is in how the equipment is used for processing. The reason the recipes that follow are processed in a pressure cooker is that the ingredients accompanying the tomatoes (which are acidy, remember) are not acidy, and must be sterilized (processed) at higher temperatures, developed by the pressure cooker or canner: 240°F. at 10 pounds pressure at sea level to 2,000 feet.

Before you start on any of the following recipes, turn back to pages 58–60 and collect the equipment and supplies described there.

Processing with a Pressure Canner

Your pressure canner may be new. In this case, read the instructions carefully, and follow them in using it.

The following twelve steps describe the method by which the pressure canner is used for pressure cooking. When used for the warm water-bath method (see page 58), the pressure gauge is not used.

With your equipment ready, and the food to be processed on hand, here are the steps you go through in pressure canning.

Step 1: Estimate how many jars you'll need for one canner load, or all the loads for the day; wash the jars thoroughly in hot, soapy water, or run them through the dishwasher. Keep them hot.

Step 2: Place the lids in a large pan and pour boiling water over them. Don't boil them, but leave them in the water until you are ready to use them.

Step 3: Store tomatoes and other vegetables or fruits in the refrigerator while they are waiting to be prepared for processing. Estimate the amount of food to be processed in one canner load and wash each piece thoroughly in several changes of cold water. (Don't let them soak, unless otherwise instructed, because water dissolves some of the minerals that give them

nutritive value.) Lift the vegetables from the washing water, then drain the water, rinsing the bowl to make sure any invisible drifts of sand are gone; repeat this washing step until the fruits or vegetables are absolutely clean.

Step 4: Prepare the vegetables for cooking per the recipe instructions. Sort the vegetables that are to be canned with a tomato recipe according to the same size and thickness. For example, pieces of celery or other vegetables of different sizes in the same jar don't look good and will cook unevenly. If there is a marked difference in size, divide the sorted vegetables so that each individual batch to be cooked will consist of similarly sized pieces. Measure the quantity that can be fitted into the number of jars the canner will handle at one time.

The right way to work your way through a bushel of vegetables is to prepare only as many vegetables as the pressure canner will process at one time. While the first batch is processing in the canner, prepare the next batch and have it ready for the pressure canner as soon as the first batch is finished.

Step 5: You now have a choice. You can decide to hot-pack the vegetables or cold-pack them. Hot-packing or cold-packing is the step that takes place *before* the cans are sealed and processed. Cold-packed vegetables are placed raw into hot jars; hot-packed vegetables are cooked briefly before packing into hot jars.

The hot-pack method ensures a better filling of the jars because air and moisture are released from the food during the heating. When you hot-pack, immerse the prepared vegetables or fruit in water that is boiling hard. Do not cover. Let the water return to a boil, and boil for the period of time directed in the recipe. Remove from the water, but do not discard the water.

Step 6: Work fast to keep the food hot. Set the funnel in the mouth of the jar and pour the tomatoes and other ingredients into the jar.

Step 7: Run a rubber spatula around the insides of the jars to remove air bubbles. Wipe the jar top and threads clean. Cap the jars as shown in the sketches on page 59. Your pressure canner is fitted with a rack. Just before the food is ready to be packed into the jars, set the pressure canner over medium heat, place the rack in the bottom of the canner, and cover with 1 inch of boiling water in small canners and 2 inches of boiling water in larger canners. As each jar is filled and the cap tightened, set it on the rack in the canner to keep hot. Pack only enough jars at one time to fill the canner. Space the jars in the canner so each receives its fair share of steam. Follow the canner instructions carefully to adjust the cover and fasten it securely. Exhaust the canner by leaving the petcock open and letting steam escape freely for seven to ten minutes. Close the petcock. *When the amount of pressure given in the recipe is showing on the pressure gauge, start counting the processing time.* Adjust the heat to keep the pressure steady.

Step 8: Process for the required length of time given in the recipe.

Step 9: When the processing time is up, remove the canner from the heat. Make no effort to lower the pressure. Let the canner stand until the pressure gauge returns to zero. Wait 2 minutes, then open the petcock slowly; if no steam escapes, the pressure is down and the cover can be removed.

Step 10: Remove the jars from the canner (use a jar lifter or a thick oven mitt) and set them upright 2 or 3 inches apart on a rack or on several thicknesses of cloth. Do not set these boiling hot jars on a cold surface or in a cold draft. Do not cover them.

Do not tighten the screw bands after processing unless so directed by the manufacturer. Cool about twelve hours.

Step 11: When the jars are cool, test the seal by pressing down on the center of the lid. If the dome is down or stays down when pressed, the jar is sealed. Remove the screw band; wash and save it.

If a jar fails to seal, repack, use a new lid, and reprocess. Or, refrigerate and use as soon as possible.

Step 12: Store in a dry, dark place with temperatures at 70°F. or somewhat below.

(Note: Process together in the canner only quarts or only pints. Don't plan to mix the two, as the processing times differ for each. Half-pints and pints require the same processing time and so can be processed together.)

Each of the recipes that follows specifies how much time the food should be boiled before processing if the hot-pack method is used.

If you are working with a canner that has no instructions, before using check the pressure gauge. Some types come equipped with a dial which gives pressure readings: This type should be returned to the manufacturer before canning season each year to make sure the dial is giving correct pressure readings. Or, the County Extension home economist may be able to refer you to a local source for checking a dial gauge.

The type of canner that has a weight gauge (no dial) can be set for pressures of 5, 10, and 15 pounds. This type of pressure gauge generally does not require checking, and a second-hand pressure canner equipped with this type of gauge generally needn't be checked for accuracy of readings.

When you are ready to pack the cooked foods into the jars, set the pressure can-

Pressure cooker with weight gauge, adjustable to 5, 10, and 15 pounds.

ner over medium heat, place the rack in the bottom and add boiling water to a depth of 1 inch for small canners, and 2 inches for very large canners. As each jar is readied and closed, set it on the rack in the canner to keep hot. Pack the canner, spacing the jars so each receives a share of steam; adjust the canner cover carefully, and fasten securely.

Allow the building steam pressure to force the air through the open petcock: steam should pour out freely for about 7 to 10 minutes. Close the petcock. When the pressure on the dial (if yours is a dial gauge) corresponds to the reading required by the recipe, start counting processing time. With the type of canner equipped with a weight gauge instead of a dial gauge, start counting processing time the minute the weight begins a rhythmic jiggling. Process for exactly the length of time called for in the recipe.

When the processing time is up, remove the canner from the heat. Don't try to lower the temperature artificially. Let the canner stand until the weighted gauge is absolutely still — or, if yours is a dial gauge, until the needle on the dial returns to zero. Open the petcock slowly. If no steam escapes, the pressure is down, and the cover can be drawn.

You are now ready to remove the jars with a jar lifter, as described in the remaining steps for boiling water-bath canning on page 62, to test for seal and so on.

Tomato Soup

Thin this soup with water or milk before serving.

24 large, ripe tomatoes (16 cups, chopped)
3 cups onions, peeled and chopped
2 cups celery, chopped
2 cups sweet peppers, stemmed and seeded (preferably red, but green will do)
1½ cups carrots, scraped and sliced
3 tsp. salt
1 tsp. dried parsley
1 tsp. soy sauce

Wash, stem, and core the tomatoes, chop them, and put through the blender ½ minute, a few cups at a time.

In a medium kettle, combine onions, celery, peppers, and carrots. Add enough water to cover, and cook until soft, about 20 minutes. Stir occasionally to keep from sticking. Press through a food mill, or blender, a little at a time.

Place the tomatoes in the kettle, add the vegetables, salt, parsley, and soy sauce, and simmer until thickened, about 1 hour.

Pour into hot jars, seal, and process 20 minutes at 10 pounds pressure at sea level to 2,000 feet. See Altitude Chart for Pressure Canning, page 59.

Yield: About 4 pints.

Vegetable and Tomato Soup

1½ quarts cold water
12 large, ripe tomatoes (8 cups, chopped)
10 cups potatoes, peeled and cubed
4 cups dried green lima beans
4 cups whole kernel corn, fresh or frozen
10 cups carrots, scraped, diced and cooked
2 cups celery stalk, cubed
2 cups onions, chopped
1 Tbs. salt
2 sprigs fresh parsley
⅛ tsp. pepper

Place the water in a large kettle over high heat. Wash, stem, core, and chop the tomatoes, and add to the kettle. Add potatoes, lima beans, corn, carrots, celery, and onions to the kettle, along with the salt, parsley, and pepper. Bring the water to a boil, and boil rapidly for 5 minutes.

Pour into hot jars, leaving 1 inch head space. Seal the jars, and process in a pressure canner for 1 hour and 25 minutes at 10 pounds pressure, at sea level to 2,000 feet.

Yield: About 7 quarts.

Celery and Tomatoes

2½ to 3½ lbs. ripe tomatoes (4 cups chopped)
4 cups celery, chopped
½ cup onion, chopped
1 tsp. salt
¼ tsp. pepper

Wash, scald, skin, stem, and core the tomatoes, and chop coarsely. Place, with the chopped celery, onion, salt, and pepper, in a medium kettle over medium heat. Bring to a rapid boil, then simmer 5 minutes.

Pour into hot jars, leaving 1 inch head space, and process in pressure canner 30 minutes, at 10 pounds pressure at sea level to 2,000 feet.

Yield: About 4 pints.

Tomatoes and Okra

2½ to 3½ lbs. ripe tomatoes (4 cups chopped)
4 cups ¼-inch okra slices
½ cup onion, chopped
1 tsp. salt
¼ tsp. thyme

Wash, scald, skin, stem, and core the tomatoes, and chop coarsely. Place in a medium kettle. Over medium heat, bring to a boil, stirring, and then simmer for 20 minutes.

Add the okra slices, the onions, salt, and thyme, and simmer 5 minutes more.

Pour into hot jars, allowing 1 inch head room. Seal the jars, and process in a pressure canner 30 minutes at 10 pounds pressure at sea level to 2,000 feet.

Yield: About 4 pints.

Meat Sauce for Spaghetti

5 lbs. ground beef
2 cups onions, peeled and chopped
1 cup green peppers, stemmed, seeded, and chopped
7 lbs. ripe tomatoes (about 9 cups chopped)
2⅔ cups tomato paste
2 Tbs. firmly packed, light brown sugar
2 Tbs. fresh parsley, minced, or 1 tablespoon, dried
1½ Tbs. salt
1 Tbs. oregano, dried
½ tsp. pepper
1 tsp. dry mustard
2 Tbs. vinegar

In a large heavy skillet over high heat, brown the meat, stirring constantly. When lightly browned, add the onions and peppers, and cook until tender.

Wash, stem, and core the tomatoes, and chop coarsely. Add to the meat, with remaining ingredients, and simmer, stirring occasionally, until the sauce has thickened, about 20 minutes. Skim away any excess fat rising to the top as a clear oily liquid.

Pour sauce into hot jars, allowing 1 inch head room. Seal, and process in pressure canner 1 hour at 10 pounds pressure at sea level, to 2,000 feet.

Yield: About 6 pints.

10
Freezing and Freezing Recipes

The fastest way out of putting by a small overload of ripe tomatoes—or green tomatoes, for that matter—is to freeze them in small batches, raw. The methods are described below.

Whole tomatoes—raw, cooked, cooked and mixed with other ingredients, as juice, as soup, in sauces and stews—freeze very well and keep without deteriorating six months or more. In other words, the recipes for putting tomatoes up by canning in the preceding chapter can be used to prepare tomato soups, stews, sauces, and vegetable dishes for freezing.

Chutneys, relishes, catsups, and the like, along with sweet tomato concoctions such as Tomato Marmalade, could be frozen, but make more sense as stored, canned products. You may prefer to allow freezer space for juice, soups, stews, and tomatoes mixed with vegetables.

Equipment for Freezing

In even the small freezer units combined with refrigerators, you can keep odds and ends of tomato leftovers, such as the Whole Tomatoes and Sliced Green Tomatoes described in the recipes below. You'll need a larger unit if you plan to freeze a lot of tomato soup, juice, and sauces. The first requisite of a freezer that will hold a lot of produce is that it maintain a temperature of 0°F. at all times: Otherwise, the produce inside, tomatoes and all, will deteriorate.

When loading your freezer, put no more food in than the freezer can freeze completely in 24 hours. This is usually 2 to 3 pounds of food per cubic foot of capacity. If you dump a whole mess of tomatoes into a small freezer, they may need several days to freeze solid, and in the process may lose their quality, a pity when they were garden-fresh to start with. Overloading can also raise the freezer temperature above the 0° point needed to keep the foods already there at top quality.

When placing tomato packages in the freezer, set them on refrigerated surfaces containing freezing coils: This guarantees rapid freezing, and helps retain the original color and texture, as well as the flavor and nutritive value of the package contents.

Packages and packaging for frozen tomatoes are as important as the canning jars are in canning. Improper packaging accounts for more (90 percent) less-than-perfect frozen produce than any other cause.

The most popular packages for freezing are plastic boxes that have very close-fitting lids, and you can stack one on the other very easily. Almost as popular, and somewhat less expensive initially, are waxed cartons meant for freezing. The cartons usually can't be reused as successfully as the plastic boxes. On the other hand, you can ruin the plastic boxes by putting them through the dishwasher with their lids: Lids almost always warp, and won't fit well again.

The types of jars called "can and freeze jars"—wide-mouthed jars with straight sides—are also used for freezing. These won't stack as readily to conserve space as the plastic boxes will.

When you are desperate for freezer containers, all sorts of hidden assets turn up on kitchen shelves: coffee cans, which can be used for freezing if they have close-fitting plastic lids, and if (only if) for tomato dishes they are first completely lined with heavy freezer-weight plastic bags. The acidy tomatoes must not come into contact with the metal sides of the containers.

Old peanut butter jars are good, and for very small batches, you can use jam or jelly jars in suitable sizes. Line the lids with freezer-weight plastic before sealing.

Packing for the Freezer

Pack tomato recipes into containers large enough to serve your family for an average meal. Thawed foods should not be refrozen except in emergencies, and even then it isn't recommended, so don't pack in containers so large you'll have to store or discard the portion you can't use immediately. A pint package is usually considered large enough for three to four portions.

Before packing tomato products in used or new containers, wash the containers and their lids thoroughly in boiling hot water, drain upside down, and allow to air dry before packing, unless you are packing a liquid, like soup.

Pack containers to within ½ inch of the rims when freezing soups, juices, stews, and sauces.

When packing individual pieces, like the Sliced Green Tomatoes, select a freezer-weight plastic bag as container, fill it with slices, gather the ends of the bag together, place your mouth over the opening and draw out all the air so that the bag gathers closely around the slices. Excluding air is done to preserve the contents.

Thawing Frozen Tomato Products

Frozen foods are usually best when thawed slowly in the refrigerator, or else at room temperature. However, you can turn the meat sauce out of a plastic freezer container, and heat it quickly over a medium heat without really harming it. Another fast-thaw approach is to place the sealed containers of frozen foods in a sinkful of cold water, or under the cold water tap turned low. Once the tomatoes in contact with the jar or box sides have thawed, the contents can be turned out of the container, and heated more quickly in a kettle.

When thawing tomato juice, just set it in the refrigerator the night before for breakfast or luncheon use.

Tomatoes Whole, Raw

Wash, scald, peel, stem and core the tomatoes. Place in heavy freezer bags, draw as much air as possible from the bag, tie it securely with a wire twist, and freeze.

To use: Take as many tomatoes from the bag as you will use, and reseal the bag after first drawing the air from it. Tomatoes frozen this way are excellent for cooking, but don't have the texture necessary for use in fresh salads.

Sliced Green Tomatoes

To freeze green tomatoes, wash, stem, and core the tomatoes, and slice into rounds ¼ inch thick. Place a square of freezer film between each tomato round. Fill heavy freezer bags with the rounds. Draw out as much air as possible for each bag, then seal with a wire twist, and freeze.

Green tomato slices are excellent in Italian dishes, or with eggs, or as a cooked vegetable. To serve as a cooked vegetable, simmer 2½–3½ pounds green tomatoes (4 cups sliced), 3 tablespoons butter or margarine, 1 teaspoon salt, ¼ teaspoon pepper, ½ to 1 teaspoon oregano until tender.

For other uses for green tomatoes, see the preceding chapter.

Index